McGraw-Hill continues to bring you the *BEST* (**B**asic **E**ngineering **S**eries and **T**ools) approach to introductory engineering education:

Bertoline, *Introduction to Graphics Communications for Engineers*

Donaldson, *The Engineering Student Survival Guide*

Eide/Jenison/Northup/Mashaw, *Introduction to Engineering Design and Problem Solving,*

Eisenberg, *A Beginner's Guide to Technical Communication*

Finkelstein, *Pocket Book of English Grammar for Engineers and Scientists*

Finkelstein, *Pocket Book of Technical Writing for Engineers and Scientists*

Gottfried, *Spreadsheet Tools for Engineers Using Excel*

Holtzapple/Reece, *Concepts in Engineering*

Palm, *Introduction to MatLab 7 for Engineers*

Pritchard, *Mathcad: A Tool for Engineering Problem Solving*

Schinzinger/Martin, *Introduction to Engineering Ethics*

Smith/Imbrie, *Teamwork and Project Management*

Tan/D'Orazio, *C Programming for Engineering and Computer Science*

Additional Titles of Interest:

Andersen, *Just Enough UNIX*

Deacon Carr/Herman/Keldsen/Miller/Wakefield, *Team Learning Assistant Workbook*

Chapman, *FORTRAN 90/95 for Scientists and Engineers*

Eide/Jenison/Northup/Mickelson, *Engineering Fundamentals and Problem Solving*

Greenlaw, *Inline/Online: Fundamentals of the Internet & the World Wide Web*

Holtzapple/Reece, *Foundations of Engineering*

Martin/Schinzinger, *Ethics in Engineering*

McGraw-Hill, *Dictionary of Engineering*

SDRC, *I-DEAS® Student Guide*

Concepts in ENGINEERING

Second Edition

Mark T. Holtzapple
W. Dan Reece

Texas A&M University

Boston   Burr Ridge, IL   Dubuque, IA   New York   San Francisco   St. Louis
Bangkok   Bogotá   Caracas   Kuala Lumpur   Lisbon   London   Madrid   Mexico City
Milan   Montreal   New Delhi Santiago   Seoul   Singapore   Sydney   Taipei   Toronto

Higher Education

CONCEPTS IN ENGINEERING, SECOND EDITION
Published by McGraw-Hill, a business unit of The McGraw-Hill Companies, Inc., 1221 Avenue of the Americas, New York, NY 10020.

Some ancillaries, including electronic and print components, may not be available to customers outside the United States.

This book is printed on acid-free paper.

1 2 3 4 5 6 7 8 9 0 DOC/DOC 0 9 8 7 6

ISBN 978–0–07–319162–1
MHID 0–07–319162–0

Senior Sponsoring Editor: *Bill Stenquist*
Outside Developmental Services: *Lachina Publishing Services*
Executive Marketing Manager: *Michael Weitz*
Senior Project Manager: *Kay J. Brimeyer*
Senior Production Supervisor: *Laura Fuller*
Associate Media Producer: *Christina Nelson*
Associate Design Coordinator: *Brenda A. Rolwes*
Cover Designer: *Studio Montage, St. Louis, Missouri*
(USE) Cover Image: *©Royalty-Free/Corbis*
Senior Photo Research Coordinator: *John Leland*
Compositor: *Lachina Publishing Services*
Typeface: *Times 10/12*
Printer: *R.R. Donnelley Crawfordsville, IN*

Library of Congress Cataloging-in-Publication Data

Holtzapple, Mark Thomas.
  Concepts in engineering / Mark T. Holtzapple, W. Dan Reece. -- 2nd ed.
    p. cm.
Includes index.
ISBN 978-0-07-319162-1 --- ISBN 0-07-319162-0 (hard copy : alk. paper)
  1. Engineering -- Textbooks. I. Reece, W. Dan. II. Title.
TA153.H65 2008
620–dc22

                        2006036342

www.mhhe.com

CONTENTS

TO THE PROFESSOR

Concepts in Engineering Second Edition introduces fundamental engineering concepts to freshman engineering students. It can be the primary text for a 1- or 2-credit introductory course that focuses on engineering fundamentals common to all engineering disciplines. Alter-natively, it can be a supporting text in a 2- or 3-credit course that teaches introductory engineering concepts and other relevant topics, such as engineering computation or graphics. The specific goals of the text follow:

- *Excite students about engineering.* We hope to stimulate students' interest in engineering by describing engineering history, challenging them with "brain teaser" problems, and explaining the creative process.
- *Build problem-solving skills.* The most important engineering skill is the ability to solve problems. We describe many heuristic approaches to creative problem solving as well as a systematic approach to solving well-defined engineering problems.
- *Cultivate professionalism.* Students are introduced to various engineering disciplines as well as other members of the technology team. They learn about the traits of successful and creative engineers. Information is provided about the advantages of obtaining advanced degrees and becoming a professional engineer. Also, students are introduced to the important topic of engineering ethics.
- *Introduce the design process.* To help freshmen experience the joy of engineering, we think it is necessary for them to work on a design project. To help support this activity, we introduce the design process.
- *Emphasize the importance of communication skills.* Too often, engineers are criticized for lacking communication skills. To help overcome this problem, we provide information on both oral and written communication that will be immediately useful to freshmen during their design project.

Because it uses only high school mathematics, *Concepts in Engineering Second Edition* can be used by students who are not calculus-ready. For students who need to review their high school mathematics, a mathematics supplement is available in printed form or at the following Web site: http://www.mhhe.com/holtzapple. It briefly describes algebra, mathematical notation, probability, geometry, trigonometry, logarithms, polynomials, zeros of equations, and calculus. Each chapter in *Concepts in Engineering Second Edition* that has mathematical content informs students of the prerequisites needed to fully understand the chapter, and directs them to the appropriate chapters in the mathematics supplement.

Concepts in Engineering Second Edition has a sister text called *Foundations of Engineering,* which provides an introduction to the following engineering science topics: computers, statistics, Newton's laws, thermodynamics, rate processes (i.e., heat transfer, fluid flow, electricity, and diffusion), statics and dynamics, and electronics. It also introduces "engineering accounting," a conceptual framework that supports engineers in all disciplines as they count the following: mass, charge, linear momentum, angular momentum, energy, entropy, and money. If you wish to include some of these topics in your course, selected chapters are available from McGraw-Hill Custom Publishing.

McGraw-Hill maintains a Web site at www.mhhe.com/holtzapple that provides supplemental teaching materials. Please visit the site; we're sure you will find it useful.

We would like to thank our reviewers for their valuable input on this edition, especially in shaping chapter 1: Lex Akers, University of Missouri; Angela R. Bielefeldt, University of Colorado; Daniel A. Gulino, Ohio University; Craig James Gunn, Michigan State University; Paul Tartaglia, Norwich University; and several others who wished to remain anonymous.

We hope that you and your students enjoy using this book. We will happily receive suggestions for improvements that may be incorporated into future editions.

Mark T. Holtzapple   W. Dan Reece

TO THE STUDENT

In a track and field race, those who are well prepared place well, and those who are poorly prepared place poorly. The purpose of *Concepts in Engineering Second Edition* is to ensure that you are well prepared, so that you will place well in the "race" you are about to begin. *Concepts in Engineering* introduces you to fundamental engineering concepts that are relevant to all engineering disciplines. The specific goals of our text follow:

- *Excite you about engineering.* Both of us are very happy that we studied engineering and enjoy using it to solve real-world problems. We hope to stimulate your interest in engineering by describing engineering history and challenging you with "brain teaser" problems. Also, we want to help develop your creativity, which is a vital part of engineering.
- *Build your problem-solving skills.* Engineers are hired to solve problems. To help you become a skilled problem solver, we describe several approaches to creative problem solving as well as a systematic approach to solving well-defined engineering problems.
- *Cultivate professionalism.* To help you choose your engineering major, we describe various engineering disciplines and how they relate to other members of the technology team, such as scientists and technicians. We also introduce you to engineering ethics and how engineers impact society.
- *Introduce the design process.* The most creative aspect of engineering is design, where we create technologies that address societal needs. To help you get started, we introduce the design process.
- *Emphasize the importance of communication skills.* Too often, engineers are criticized for lacking communication skills. To overcome this problem, we provide information on both oral and written communication that will be useful throughout your career.

Engineers must master mathematics. If you need to review your high school mathematics, a mathematics supplement is available at the following Web site: http://www.mhhe.com/holtzapple. It briefly describes algebra, mathematical notation, probability, geometry, trigonometry, logarithms, polynomials, zeros of equations, and calculus. Each chapter in *Concepts in Engineering Second Edition* that has mathematical content informs you of the prerequisites needed to fully understand the chapter, and directs you to the appropriate chapters in the mathematics supplement.

McGraw-Hill maintains a Web site at www.mhhe.com/holtzapple that provides supplemental materials. Please visit the site; we're sure you will find it useful.

We hope that you enjoy using this book. We will happily receive suggestions for improvements that may be incorporated into future editions.

Mark T. Holtzapple   W. Dan Reece

ABOUT THE AUTHORS

Mark T. Holtzapple

Mark T. Holtzapple is Professor of Chemical Engineering at Texas A&M University. In 1978, he received his BS in chemical engineering from Cornell University. In 1981, he received his PhD from the University of Pennsylvania. His PhD research focused on developing a process to convert fast-growing poplar trees into ethanol fuel.

After completing his formal education, in 1981 Mark joined the U.S. Army and helped develop a portable backpack cooling device to alleviate heat stress in soldiers wearing chemical protective clothing.

After completing his military service, in 1986 Mark joined the Department of Chemical Engineering at Texas A&M University. It quickly became apparent that he had a passion for teaching: within a 2-year period he won nearly every major teaching award offered at Texas A&M, including the Tenneco Meritorious Teaching Award, the General Dynamics Excellence in Teaching Award, the Dow Excellence in Teaching Award, and two awards offered by the Texas A&M Association of Former Students. Mark particularly has a passion for teaching freshman engineering students. He wrote this book to excite students about engineering.

In addition to his role as an educator, Mark is a prolific inventor. He is developing an energy-efficient, ecologically friendly air-conditioning system that uses water instead of Freons as the working fluid. He is also developing a high-efficiency, low-pollution Brayton cycle engine suitable for automotive use. In addition, he is developing technologies for converting waste biomass into useful products, such as animal feeds, industrial chemicals, and fuels. To recognize his contributions in biomass conversion, in 1996 he received the Presidential Green Chemistry Challenge Award offered by the president and vice president of the United States.

W. Dan Reece

Dr. Reece is a Professor in the Nuclear Engineering Department and Director of the Nuclear Science Center at Texas A&M University. He received his Bachelor of Chemical Engineering, Master of Science in Nuclear Engineering, and PhD in Mechanical Engineering, all at the Georgia Institute of Technology. He has worked as an analytical chemist, a chemical engineer, and a staff scientist at the Pacific Northwest National Laboratory, before his current positions at Texas A&M.

Much of Dr. Reece's research is in the area of radiation monitoring, novel uses of radiation in medicine, and the health effects of radiation. Like Dr. Holtzapple, he has a passion for teaching and has won a Distinguished Teaching Award from the Texas A&M Association of Former Students. Dr. Reece teaches many topical courses in dosimetry and health physics, has an active consulting business, and, whenever his schedule allows him free time, enjoys backpacking, playing tennis, and running. His greatest enjoyment comes from his children, his students, and the advances in medicine and worker protection he has helped to make.

CHAPTER 1

Preparing to Be an Engineer

Toto, I have a feeling we're not in Kansas anymore.
—Dorothy Gale in The Wizard of Oz

When the tornado flew Dorothy to the Land of Oz, she quickly realized that her world had changed. No longer was she in the safe comfort of her family in Kansas; rather, she was in a strange land where everything was different. As a beginning engineering student, you may be able to relate to Dorothy—your world has just changed, too.

In this chapter, we would like to examine the changes in your life, some apparent and some subtle. After previewing some of those changes, we will help you think about where you would like to go during the next few years. Finally, we'll give you some hints on how you might get there and point to some guideposts along the way.

1.1   What Have I Done?

In many ways, college is not Kansas anymore. When you were in high school, you surely were among the brightest students in the school. In engineering college, everyone is bright; perhaps you will no longer be among the top students. (Remember, 50% of all engineers graduated in the bottom half of their class.) When you were in high school, your parents likely told you that it was the most exciting time of your life. Then when you graduated, they said that college would be the most exciting time of your life. Well, they were right. It really is different here, one of the most exciting (and challenging) times of your life.

For a moment, let us be your favorite uncles and give you a quick look at how college differs from high school. First and foremost, your parents aren't here. Because they no longer regulate your day-to-day life, now *you* must monitor your own behavior. This is not as easy as it seems—there are many more diversions and distractions at college than in high school. These diversions and distractions usually arrive in the form of a fellow student saying, "Hey! Do you know what's going on at the Student Center today?" Or, "Are you taking a date to the concert tomorrow night?" Or any one of an endless list of other temptations.

We aren't suggesting that you shouldn't sample these various diversions; however, as with many things, it's an issue of balance. For example, there's nothing wrong with eating a piece of chocolate cake now and then, but there's plenty wrong with eating seven chocolate cakes all at once. Regarding academic distractions, how many of these diversions can you take before it affects your studies? Most of your college learning will occur in the midst of these diversions, so can you stay focused on your goal of becoming an engineer? Meeting people with different ideas and being exposed to new activities will enrich your life, but will they divert you from your ultimate objective? A common theme in this opening chapter is that you must find balance among so many choices.

Stated another way, to succeed in college, you must behave like—*gasp!*—an adult. There's a long list of skills that are necessary to pass into adulthood. For those who are sure that you are already grown up, consider grading yourself on the following partial list.

1.1.1 You Must Develop Basic Living Skills

Unless you went to a boarding school, your parents have probably provided basic services, such as keeping you in clean clothes, washed, fed, and moved from place to place. Now that you are on your own, this is no longer a given. You must now provide these basic services for yourself.

Zits, by Jerry Scott and Jim Borgman

1.1.2 You Must Overcome Defeats and Obstacles More on Your Own

College is tough, and engineering is among the toughest curricula in college. You will be challenged continuously. College is boot camp for the mind. You will fail an exam. Labs won't work. Class projects will go astray. You'll have four finals in two days. There will be a class that is simply incomprehensible. Your teammates in your senior design class will seem to have been asleep for the last four years.

In your personal life there will also be bad days. Your guy (or girl) will dump you for no particular reason. It will rain for three weeks straight. Your roommate will steal your food. Your dorm mates will make so much noise you cannot sleep at night.

Your family won't be there to help you pick up the pieces. Your friends may be busy or going through some of the same challenges that you are. These are the times that will try your soul, and you'll have to reach deep and pull yourself up.

On the other hand, most students *do* survive this process; it's just part of growing up. And, unlike boot camp, you receive time off for good behavior. It's also fair to say that, along with all these challenges and obstacles, some of the best hours in your life will occur in college.

1.1.3 You Must Manage Your Time

Professors will sling information at you faster than you've ever seen before (imagine drinking water from a fire hose). You'll find that you are challenged as never before to get everything done.

Let's suppose that you're taking 15 hours of classes, and that these hours include six hours of lab. Because you want to do well, you spend three hours studying for every hour in class. That's 60 hours per week of your time that is already gone, but you still must sleep, eat, do laundry, and, we hope, have a bit of fun. Working within that schedule is a challenge. A bit later we'll talk about how you can accomplish this and survive.

1.1.4 You Must Manage Your Money

College is expensive. Textbooks are expensive. Housing is expensive. Cars are expensive. You get the idea. Unless you're very lucky and have rich parents, you not only must budget your time but also your money. Some people have natural money-management skills; others don't. There are many strategies for managing money, but perhaps the simplest is *divide and conquer.* If your money comes in lumps, resist the temptation to spend it while you've got it. Divide the lump into weekly expenses, and if some remains, then you have "mad money" for the extras in life. Track expenses. After a semester, you'll know what expenses come when. This helps you and your money source—likely some relative— know how much and when you will need it. Get help managing what you've got. If you need help, you won't be the first one. Finally, look around to see if there are ways to earn more money. There are many opportunities—from scholarships to jobs with flexible hours—that can help keep your wallet from being empty.

If you decide to take a part-time job, be careful that the job does not slow your graduation. Student wages are fairly low, so you must weigh the value of money now against the much greater amount of money you could earn by graduating more quickly. If the part-time job hurts your grades, it could reduce your job offers upon graduation. If you are in a financial crunch, you might consider taking student loans or taking a semester off to earn money. If you must take a part-time job while a full-time student, limit your employment time to 10 hours per week.

1.1.5 You Must Get Along with People

It sounds funny giving this advice to engineering students, but it's true. Outwardly, it appears that engineers don't need people skills because they focus on science, mathematics, and technology. In reality, to be successful, you must work well with other people. Engineering students work in study groups, and practicing engineers work in engineering teams. The socialization skills that you learn in college will help you tremendously when you go to work. Once you leave college and take your first job, you'll probably find that you work in teams with people who are not engineers. It will be imperative that you get along and can help the team achieve its goals.

Zits, by Jerry Scott and Jim Borgman

1.1.6 You Must Stay Healthy

While you are away from home for the first time, you must manage what you eat, get enough exercise, sleep, stay healthy, and avoid the "freshmen 15" (the 15 pounds the average freshman gains during the first year of college).

In the middle of all these professors beating on your brain, you must take care of your body, too. Remember all the nagging your mom gave you about eating well-balanced, nutritious meals that include vegetables? Well, she was right, and we'll tell you the same. Between meals, if your blood sugar drops, avoid snacking on sugary foods. Soon after eating the sugar, your blood sugar will be replenished, and you'll feel good again. However, in response to the slug of sugar, the body secretes insulin, which causes sugar to be taken up by cells. The end result is that your blood sugar levels can crash, making you feel worse than before you ate the snack. Instead, snack on foods that slowly release sugar, like apples or nuts.

Zits, by Jerry Scott and Jim Borgman

Here at the beginning of your adult life, your habits will start a pattern that will either help you or hurt you. Because college is boot camp for the brain, you'll have to make time to maintain your exercise routine. It's important. Fortunately, you'll find plenty of people who love to play your favorite sport or will show you a new one. Exercise will help you sleep, help you stay alert during those long afternoon classes, and make you feel better.

Be sure to find time for sleep. Getting the proper amount of sleep ensures that your study time is productive. To prepare for exams or complete projects, students sometimes pull all-nighters. Although an all-nighter may get you through the latest crisis, it sets you up for more crises: The next day you are exhausted and it will be hard to concentrate on your other classes. Some students take caffeine to stay awake. If taken late at night, caffeine can actually disrupt your sleep patterns, which makes you drowsy the next day, so you need more caffeine. Intermittent use of caffeine can cause withdrawal symptoms, such as headaches. Although caffeine is generally regarded as safe, if you use it, be cautious and thoughtful.

On the bright side, with all these new responsibilities comes a great gift: You may choose which opportunities you want to explore, which friends you have, what paths to walk. It's here in college that you begin to figure out who you really are, and with each success, with each obstacle overcome, you gain confidence and a sense of identity.

1.2   The Classroom . . . It's Not Kansas Either

Education is what is left over after you have forgotten everything you learned in college.
—Robert Coles

In college, you'll find that your academic environment has changed a lot. By law, anyone can go high school. However, attending college is a privilege, and, as you well know, you must demonstrate above-average intelligence and scholastic achievement to get in. By definition, the people in college are very smart. If you're used to being the smartest kid in your class—or at least in the top 25%—things may be very different here. One common problem is that bright high school students, who could be among the best without investing much effort, are often just average in the college population. Some students adjust to this, rally, move up to the next level, and take their place among the college best. In contrast, some are caught off guard and don't know how to proceed.

Calvin & Hobbes, by Bill Watterson

On this note, let us say that your high school study habits won't carry you through college. You must change your learning strategies and become a more efficient learner. In college, material is presented much faster than in high school. Furthermore, you'll be expected to learn much of the material on your own. Finally, you will be expected not to simply learn facts, but to synthesize the material and demonstrate your knowledge at a much higher level.

Also notice that the classroom atmosphere has changed, too. Generally, your professors are indifferent; they probably won't hold it against you if you don't come to class, don't turn in your homework, or fail an exam. It's not that they don't care; it's part of the adult behavior they expect of you. You're not here because you have to be; you're here for *you*.

Calvin & Hobbes, by Bill Watterson

Speaking of professors, you should get to know them. Professors provide office hours where you can meet with them one on one. Office hours are particularly useful if you have trouble with a homework problem. As you get to know professors, you may find that you want to become involved with their research projects. If you like doing research at the undergraduate level, you may decide that you want to go to graduate school. By getting to know professors, they can provide you with letters of recommendation for graduate school or your first industry job.

1.3   The Emerald City—Where Is It You're Going, Again?

You got to be careful if you don't know where you're going, because you might not get there.
—*Yogi Berra*

Let's briefly discuss your overall goal: receiving an engineering degree. We believe it is important for you to understand your motivation, because, as we have already hinted, you'll be tested from time to time. Let us say right now that engineering is the most wonderful job in the world. As an engineer, you'll probably understand more about our day-to-day world than anyone else in any other profession. It's a remarkably creative profession; you will spend your working life creating things that have not existed before. Almost always your work will help other people. Engineering offers a variety of problems and challenges second to no other occupation. For most engineers, their jobs next year will be very different from what they're doing now. If you want to change and move to another company, your skills give you mobility that few others have. You will be paid well, but that's not really the point (although money certainly opens many doors). What matters most is that you can do enjoyable and important work surrounded by smart people.

It is worth spending a little time thinking about why you are studying engineering. During our time as professors, we have gleaned the following motivations from various students:

- My parents said they would pay my tuition only if I studied engineering.
- I am good in math and science, so my guidance counselor said I should be an engineer.
- Uncle Fred is an engineer. He makes lots of money. I want to be like Uncle Fred.
- I heard that engineering is hard. I want to test my abilities, so I picked engineering.
- I am a creative person who loves technology.
- I have a passion for solving practical problems that can make a difference in the world.

Your motivations may be similar to some of those above, or perhaps you have your own. Whatever your motivations, you will be most successful if they are internally generated. It is hard to perform well if you are simply trying to please others or just want to make money. Although high salaries are certainly a great benefit of engineering, we believe that money alone is a poor motivator. If all you really want is money, go to business school and learn to make deals.

Some of you will wonder, partway through your academic career, if engineering or your particular engineering discipline is really the right choice for you. Part of this is normal: Everyone going through boot camp has doubts from time to time. Then again, this may be something real. Significant numbers of undergraduate engineering students change majors, usually to another engineering discipline. How can you find out what's right for you?

One way is to test the waters with internships, with mentor groups, and especially by joining student chapters of engineering professional societies. This will give you the opportunity to look ahead and see what other people are doing in your field so you can see if it fits. It's a great opportunity to look into the future, so its importance cannot be overemphasized. If you're thinking of changing to another engineering discipline, attend their chapter meetings to see if this new discipline fits you better.

Internships and co-op positions are golden opportunities. Through these programs, juniors and seniors (sometimes freshmen and sophomores) can often get a job in their field. You can see what engineers in your profession actually do for a living, and you are paid well. If the company likes you, there may be a job waiting for you when you graduate. Can it get any better? If you remember our talk about managing money, here is a way you can earn more money while learning what it is you're aiming for. If there's any way you can make this happen, do it.

1.4 The Yellow Brick Road

We've talked a bit about the college environment and we've expounded about your future, so now let's discuss how you can get there.

1.4.1 Study Skills

Think for a moment about learning itself. How does learning occur? The short answer: You rewire neurons in your brain. It's not our lectures, it's not your textbook, and it's not your classroom activities. Although all these things are important, they're just tools or environments; it's really just you and your brain. If you think about it, we professors will see you for only about 40 hours during the semester. Do you really think during that short time we can teach you everything you need to know about thermodynamics, for example? We can't rewire your neurons; only you can.

With that in mind, let's think for a moment about how you learn. There are literally hundreds of books about learning and cognition, but in reality it's up to you. *You* must figure out the best way for *you* to learn. Some people learn best by the spoken word; lectures are very important for them. Some learn best visually; pictures and the written word are best for them. Some learn best by hands-on experience; labs and projects fit them best. In truth, all of us need to be proficient in all these ways, and each has its own benefits. Find out what works best for you and emphasize that.

The world's knowledge has grown large indeed. Your professors have so little time to present things to you that we often speak in generalities—it's the only way to possibly cover as much as we would like. However, this puts a particular burden on you. Exams usually ask particulars, not generalities. It's up to you to learn how to apply generalities to the particulars in preparation for your tests.

Let's give you a funny example. Do you know the names of the months of year? All twelve of them? Are you sure? Really well? OK. Now for the test: Time yourself and give them to us alphabetically. Why was that so hard? Because it's not the way you learned them. It's the same way with going from generalities to specific problems.

Knowing the importance of being able to go from the general to the specific, here's our advice: Approach homework with determined enthusiasm. Here is your chance to practice applying generalities to particular problems. Certainly work all the problems assigned, but also work some of the others. Work old exams (if you can get them), not just by looking at the answers if they're available, but work them out yourself. It is only with a lot of practice that you'll become good at this. Another great gift from homework is that you find the holes in your knowledge. Even better, you'll find the parts of your understanding that are wrong. The hardest weed to root out is errors in understanding; when you have it wrong, it's invisible until you stumble over it.

Now let's get down to specifics on how best to rewire those neurons.

Create a Good Environment

Many studies show that your study environment greatly affects the speed at which you can master material. Be sure the room is not too hot; it will make you sleepy. Don't lie in bed to study; use a straight-backed chair at a desk. Be sure there is adequate lighting. Quiet is good. Distractions, even minor ones, slow down the learning process. With practice one can learn to focus and ignore many of these distractions, but they still slow the overall learning process. You may wish to play quiet music (no words, instrumental only) in the background to help drown out other noises. Some researchers claim that playing Mozart has particular benefits to learning, which is dubbed the "Mozart effect." Some students can

Calvin & Hobbes, by Bill Watterson

Calvin & Hobbes, by Bill Watterson

Calvin & Hobbes, by Bill Watterson

use the library or other quiet place and do well. Some students need the support of familiar and comfortable surroundings. Find what works best for you, but remember: Distractions are your enemy.

Arrive at Class Prepared

As we mentioned above, you won't see your professor very much. And the lecture material is presented only once. Your best strategy is to read the assigned material covered in lecture ahead of time, allowing you to make optimal use of each lecture. If you arrive already familiar with the material, you can see what your professor emphasizes; this can be a valuable clue as to what will be on the exam. If there are things you don't understand, or the lecture seems different from the material you read, you can ask questions. This is the best time to get things straight, right at the beginning.

Take Good Notes

Because you've read the material, it should be fairly easy to take notes. Take the best notes you can. Try to understand the points being made as you take notes. If you fall behind, write down those points you missed, and fill in later from the material in the textbook. Soon after class, review your notes and fill in any holes. This makes sure that your notes are complete, and this is your first review of the material. Only through repetition can you understand and retain the information given. If you have questions, you have time to get them answered before it becomes an emergency. It's also much easier to learn it right the first time rather than try to correct what you think you know later.

Get the Most out of Your Textbooks

Although textbooks make wonderful paperweights, this is not their true purpose. Your textbook expands upon the lecture material, fills in gaps in understanding, provides a different perspective, gives examples, and helps you truly master the material. You need to read it before class, read it after taking your notes, and scan it again before you take your exams.

As an aside, let us be very clear by what we mean by "reading" your text. We don't mean reading as you do with the newspaper or with a novel. (Can you remember the details of a newspaper article from a week ago?) Reading a textbook requires full attention; remember, you are rewiring your neurons. Use a highlighter; you might want to use one color highlighter to mark major words or concepts and a different color for subordinate information, but don't highlight the whole page. Write in the margins. Make notes to yourself summarizing what you do understand, and a list of questions for those things that you don't. Take the question list with you when you seek help so you won't forget to cover everything. By the time you complete your course, your textbook will be customized for you, and will make a great addition to your engineering library. (Engineers are hired for their knowledge. Most of us have awful memories, so we keep an engineering library to help us remember the details.)

Work Many Problems

Homework is your best chance to practice before taking an exam. It is your best opportunity to practice applying general principles to specific problems. Without diligent work, you cannot succeed nor can you master the material. Homework *is* diligent work. To do homework, you will review your notes and consult your textbook; through this process, you will gain a more complete understanding. If there's only one thing you do well in college, let it be homework. A final word of advice: When you're stuck, seek help. This doesn't mean you're stupid. It doesn't mean you are weak or lacking. It means you're smart enough to spend your time wisely. Everyone needs help sometime during college. Don't wait until it's too late to find help.

Use Repetition Again

Test time is coming. Good news! You're pretty well prepared. You've read the text before class, you took notes during class, and you reviewed and completed your notes after class. You read your textbook again, you've worked problems, and now you're almost there. The last step—perhaps most important—is to go over the material one more time. Repetition may be boring, but it's how you rewire the neurons and keep them rewired. There is no other way.

Repetition is so important to learning that topics are often repeated year after year. Just because you were exposed to Newton's laws in high school does not mean you will never see this topic again. To the contrary, you will revisit many topics throughout your career.

Figure 1.1 shows the spiral of learning. In your freshman year, you will learn topics at a level appropriate for a freshman. In your sophomore year, you will repeat many of these topics, but at a level appropriate for a sophomore. This process of revisiting a topic, but in more depth each time, will continue through your whole career.

1.4.2 Master Exams

Some people do well under pressure, and exams bring out the best in them. The rest of us are not so fortunate. If the stress of an exam makes you lock up, practice taking exams beforehand. Choose several problems out of your textbook and see if you can work them in a specified time. Work old exams. We promise you will become better with practice.

Try to figure out what your professor will ask. If you paid attention during class while taking notes and you analyzed which problems were assigned for homework, you'll already be ahead of the game. Students often have access to old exams, which are a great resource. Working through them will give you confidence or motivate you to study more.

During the exam, skim over it first to familiarize yourself with it. Then answer the easiest questions. If you get stuck, move on—your subconscious can work on the hard problems while you continue working on the easier questions. Don't give up, and use the full time allowed.

1.4.3 Time Management Skills

> *Simplify, simplify, simplify.*
> *—Henry David Thoreau*

The best way to simplify your life is to get organized; a day planner is one of the best tools for accomplishing that.

Let's look at how your planner might be customized. First, set out the classes and labs. Second, set the times for preparing for class and for completing and reviewing your class notes. Third, take care of yourself. Set times for meals, laundry, and the other necessities of life. Set down times to review notes, read your textbook, and do homework. Finally, block out free time for recharging your batteries and exploring the many opportunities that campus life offers you.

If you look at the daunting schedule in Figure 1.2, it's a visual clue of how busy you're going to be. Even with all this planning, you'll have to be flexible. Exams happen—although, with your preparation, tests won't be the trial that many students experience by trying to learn everything the night before. Nevertheless, tests still require time for preparation.

As you plan your study schedule, remember that 50 minutes is about the length of time that you can focus and remain productive. Allow yourself some breaks so that you

FIGURE 1.1
The Spiral of Learning

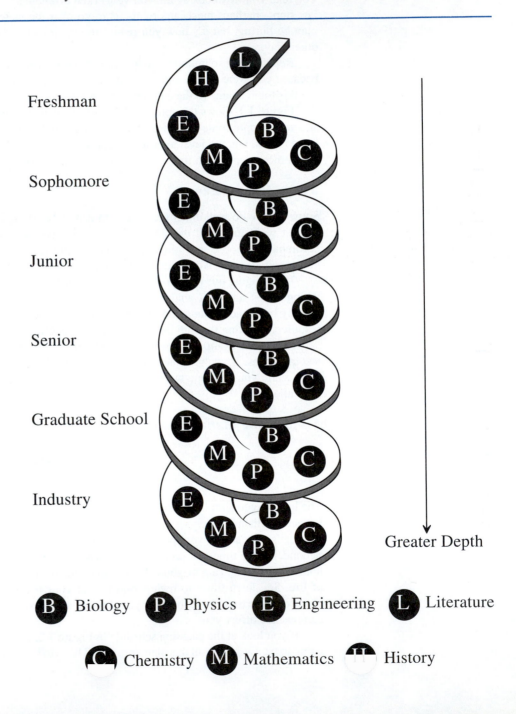

	Sunday	Monday	Tuesday	Wednesday	Thursday	Friday	Saturday
5:00							
6:00							
7:00		breakfast	breakfast	breakfast	breakfast	breakfast	
8:00	breakfast	physics prep	physics homework	physics prep	physics homework	physics prep	
9:00	math homework	English		English		English	breakfast
10:00		physics	engineering prep	physics	engineering prep	physics	study for math test
11:00	church	review notes	engineering	review notes	engineering	review notes	
12:00	lunch	lunch	lunch	lunch	lunch	lunch	lunch
1:00	call home	math prep	review notes	math prep	review notes	math prep	football game
2:00	shopping	math	math prep	math	physics lab	math	
3:00		review notes	math lab	review notes		review notes	
4:00	laundry	gym		gym		gym	
5:00	supper	supper	supper	supper	supper	supper	out for supper
6:00	math homework	engineering homework	physics homework	engineering project	write physics lab report	out to play	
7:00							
8:00	English reading						study for math test
9:00		math homework	English reading	English homework	English reading		
10:00							
11:00							

FIGURE 1.2
Customized Day Planner Week

come back to your studies refreshed. The breaks could involve socializing with friends, watching television, or even doing laundry. (Trust us: As a study break, you will actually look forward to doing laundry.) Be flexible. If you are exhausted at some particular time, it may be wise to use this time to go to a movie and study later during scheduled free time.

Limit distractions: computer games, television, or talkative roommates. Some of these can be mastered with technology. If you have a TV show you can't live without, record it to watch when it better suits your needs. Voicemail is a great gift. Simple appliances, such as a microwave (if your dormitory regulations allow it), can help you multitask—study while heating food, for example.

Your living arrangements can affect your time management. For example, if you live in an apartment, you will be responsible for purchasing groceries and cooking, both of which consume large amounts of time. In contrast, if you live in a dormitory and eat at a dining hall, you will save this time for studies. If you live on (or near) campus, your commuting time is minimized. If you live a significant distance from campus, the time you spend commuting will cut into your study time.

Two pieces of advice concerning computer technology: Save copies! Back up often! Those who live by the computer can die by the computer. Everyone we know (including us!) has a horror story about an assignment, thesis, or journal article being lost in cyberspace.

1.4.4 Teamwork

Your fellow students are a valuable resource: Some will help you through homework sets or a class that is seemingly incomprehensible. Their questions will often illuminate the holes in your knowledge. Practice grace under pressure with your classmates and teammates. These skills are valuable throughout life. Watch those who work well with groups and analyze how they do it. It is a skill that can be learned.

1.4.5 Use Resources

Your college abounds with resources that can help you. Early in your academic career, locate and take advantage of the following.

- *Mentors.* Most colleges have a way to pair you with a mentor, either an upperclassman or someone from industry in your field. The opportunities here are huge. Upperclassman can often give invaluable advice on professors and courses, furnish old exams, and otherwise be the helpful older sibling you always wished you'd had. The mentor from industry won't be able to help you much in day-to-day skirmishes, but can advise what courses might be useful when you finally graduate, can open doors to internships, and can be helpful when you wonder what you want to be when you grow up.
- *Tutors.* These helpful people range from paid professionals to some of your brighter friends. The object of the time spent at the university is to master the material, not to be a martyr. Everyone needs help somewhere along the way. You won't be different. Our advice? Don't wait too long. Sooner is often much, much better than later. Honor societies often furnish tutors to undergraduates for free.
- *Books.* Read your textbooks as if they determine your fate. (In a way, they do.) Often, supplemental texts will be suggested. Look through them. Sometimes you will be lucky and find where your professor gets test questions! If your assigned textbook is seemingly written in Martian, maybe a supplemental text will be written in a way that is easier for you to understand.
- *Libraries.* Few places other than the classroom offer so much. Knowledge is amassed, arranged, and sorted. What more could you ask? And, yes, it's also quiet and usually one of the best places to study. Be respectful of the library, and use it often.
- *TAs.* Teaching assistants are there to help you. If they seem distracted, remember that they have their own studies, so be patient. However, insist that they do their job, which is to help you learn. A kind word from you when they help will be unexpected and appreciated.
- *Professors.* Professors come in many varieties, so make allowances. They are tremendous sources of knowledge, and they write your tests. Treat them with respect (for

example, don't ask them to explain the last six weeks of lectures the day before the exam), and you will find that most are truly on your side.

This list is not exhaustive. Keep looking. The department of student affairs or the academic office can often steer you to needed help.

1.5   Some Final Thoughts

Here are additional practices that will help you succeed in college and life.

- *Network.* The folks you're going to school with now will, like you, graduate and scatter to the four winds. Thus, you will have friends and contacts all over the world. Someday, say you decide to change jobs, these contacts will be golden.
- *Explore.* In the next four to five years, you will be exposed to more ideas than you can imagine. Some will be instructive and lead to interesting discussions and perhaps change the way you look at things. Some could have the potential to do tremendous and permanent damage. Be open, but also be circumspect.
- *Learn to write well.* We can't think of better advice. Writing well and being able to present your ideas well is a skill that will carry you far throughout your life. The best book on writing that we've found is *Style: Toward Clarity and Grace* by Joseph M. Williams (University of Chicago Press, 1995). Get it. Read it. Use what it says.
- *Learn to reason well.* As an engineer, you will have many intricate puzzles to solve. Unfortunately, Mother Nature is unforgiving and does not tolerate mistakes. Only with practice will your reasoning become better and clearer. If you make a mistake, with practice you will more quickly find your errors in logic or assumptions.
- *Distinguish grades from learning.* Grades are certainly important. They provide feedback on how well you are learning the material, and they open doors for future jobs or schooling. Nonetheless, grades are largely determined by exams, which are artificial. (We've never heard an employer say, "Here, solve this problem in five minutes or you're fired.") Grades measure only a portion of the skills you need to be successful in life. Mark Twain emphasized this point by saying, "Never let mere schooling get in the way of your education." While you are striving for good grades, be sure to learn those things not measured by grades.
- *Be well rounded.* Engineers are amazingly well-rounded individuals. Some are musicians; others are history buffs, or pilots, or photographers, or athletes, or. . . . Keep pursuing your outside interests. They will keep you engaged with life, plus they lead to way more delightful dinnertime conversation than discussions of heat transfer coefficients.
- *Read.* Numerous magazines (e.g., *Popular Science, Scientific American, Science News, Science*) can keep you updated on science and technology. Read them for fun.
- *Plan your first job interview.* Although your first job interview may be four to five years away, what are you going to talk about that would be compelling to a potential employer? Typical job interviews last about 30 minutes. A discussion of your grades and favorite classes may take up five minutes. Then what?

- *Distinguish yourself from other engineers.* Regardless of which college they attend, engineering students have a similar curriculum within each engineering discipline. When you graduate, there will be literally thousands of engineers with formal training similar to yours. How will you distinguish yourself from these other engineers?

- *Make a thousand friends.* Here, away from your family and all the folks who have known you since birth, is the place where you begin to discover who you really are. Friends: You can't have too many, but choose them well.

- *Take time for dating.* College is a rare opportunity to be with thousands of single people your own age. You will find that you have much in common with many of them. Take time to find someone with whom you might like to spend the rest of your life.

- *Exercise iron discipline.* Is it possible to follow all the advice given here all the time? Probably not. But we have outlined a true path, and the more often your feet follow this path, the better student and the better engineer you will be. If you fail gloriously sometime, take it as a lesson, pick yourself up, learn from it, and move forward. No more is required.

1.6 Summary

We've tried to present our ideas here succinctly and with humor; we hope it has been fun to read. However humorous our words have been, the thoughts we've tried to convey are serious. We find students fail most often not because they lacked some specific skill or native intelligence, but simply because they didn't understand what college was all about. In these few pages, we've tried to give you an introduction. Many other resources can help you along your way, including counselors, survival guides, student organizations, professional tutors, your friends, and fellow classmates. We wish you the best of luck and remind you to have fun in the midst of all your challenges at college.

Further Reading

Donaldson, K. *The Engineering Student Survival Guide,* 3rd ed. New York: McGraw-Hill, 2004.
Landis, R. B. *Studying Engineering: A Road Map to a Rewarding Career,* 2nd ed. Los Angeles: Discovery Press, 2000.

PROBLEMS

1.1 Prepare a list of your short-range (one-year), intermediate-range (five-year), and long-range (20-year) goals.

1.2 Do a self-inventory and list your strengths and weaknesses. Devise a plan for overcoming your weaknesses.

1.3 Do a self-inventory and list activities that interest you.

1.4 Do a self-inventory and list those things about which you are passionate.

1.5 Do a self-inventory and list accomplishments in life that would make you happy.

1.6 Prepare a weekly schedule of your activities, including classes, study time, sleep, exercise, eating, and so on.

1.7 Write a brief report on the Mozart effect.

1.8 Write a brief report on foods that boost learning, so-called brain foods.

1.9 Write a brief report on the effects of lack of sleep.

1.10 Write a brief report on the health benefits of exercise.

1.11 Write a brief report on the latest thinking about a healthy diet.

1.12 Prepare a list of traits that you find attractive in other people. Identify those traits that you find in yourself.

1.13 Take a walking tour of your university to learn the location of key services (e.g., your professors' offices, your academic advisor's office, student mental health services, student medical center, library, dean's office, co-operative education office, student housing office, job placement office). Write a brief report summarizing their function and location.

CHAPTER 2

The Engineer

Nearly all the manmade objects that surround you result from the efforts of engineers. Just think of all that went into making the chair upon which you sit. Its metal components came from ores extracted from mines designed by mining engineers. The metal ores were refined by metallurgical engineers in mills that civil and mechanical engineers helped build. Mechanical engineers designed the chair components as well as the machines that fabricated them. The polymers and fabrics in the chair were probably derived from oil that was produced by petroleum engineers and refined by chemical engineers. The assembled chair was delivered to you in a truck that was designed by mechanical, aerospace, and electrical engineers, in plants that industrial engineers optimized to make best use of space, capital, and labor. The roads on which the truck traveled were designed and constructed by civil engineers.

Obviously, engineers play an important role in bringing ordinary objects to market. In addition, engineers are key players in some of the most exciting ventures of humankind. For example, the Apollo program was a wonderful enterprise in which humankind was freed from the confinement of earth and landed on the moon. It was an engineering achievement that captivated the United States and the world. Some pundits say the astronauts never should have gone to the moon, simply because all other achievements pale in comparison; however, we say that even more exciting challenges await you and your generation.

2.1 WHAT IS AN ENGINEER?

Engineers are individuals who combine knowledge of science, mathematics, and economics to solve technical problems that confront society. It is our practical knowledge that distinguishes engineers from scientists, for they too are masters of science and mathematics. Our emphasis on the practical was eloquently stated by the engineer A. M. Wellington (1847–1895), who described engineering as "the art of doing . . . well with one dollar, which any bungler can do with two."

Although engineers must be very cost-conscious when making ordinary objects for consumer use, some engineering projects are not governed strictly by cost considerations. President Kennedy promised the world that the Apollo program would place a man on the moon prior to 1970. Our national reputation was at stake and we were trying to prove our technical prowess to the Soviet Union in space, rather than on the battlefield. Cost was a secondary consideration; landing on the moon was the primary consideration. Thus, engineers can be viewed as problem solvers who assemble the necessary resources to achieve a clearly defined technical objective.

Engineer: Origins of the Word

The root of the word *engineer* derives from *engine* and *ingenious*, both of which come from the Latin root *in generare*, meaning "to create." In early English, the verb *engine* meant "to contrive" or "to create."

The word *engineer* traces to around A.D. 200, when the Christian author Tertullian described a Roman attack on the Carthaginians using a battering ram described by him as an *ingenium*, an ingenious invention. Later, around A.D. 1200, a person responsible for developing such ingenious engines of war (battering rams, floating bridges, assault towers, catapults, etc.) was dubbed an *ingeniator*. In the 1500s, as the meaning of "engines" was broadened, an engineer was a person who made engines. Today, we would classify a builder of engines as a mechanical engineer, because an engineer, in the more general sense, is "a person who applies science, mathematics, and economics to meet the needs of humankind."

2.2   THE ENGINEER AS PROBLEM SOLVER

Engineers are problem solvers. Given the historical roots of the word *engineer* (see box above), we can expand this to say that engineers are *ingenious* problem solvers.

Photo courtesy of NASA/NASA Media Services.

Fulfilling President Kennedy's promise, the United States landed on the moon in 1969.

In a sense, all humans are engineers. A child playing with building blocks who learns how to construct a taller structure is doing engineering. A secretary who stabilizes a wobbly desk by inserting a piece of cardboard under the short leg has engineered a solution to the problem.

Early in human history, there were no formal schools to teach engineering. Engineering was performed by those who had a gift for manipulating the physical world to achieve a practical goal. Often, it would be learned through apprenticeship with experienced practitioners. This approach resulted in some remarkable accomplishments. Appendix C summarizes some outstanding engineering feats of the past.

Current engineering education emphasizes mathematics, science, and economics, making engineering an "applied science." Historically, this was not true; rather, engineers were largely guided by intuition and experience gained either personally or vicariously. For example, many great buildings, aqueducts, tunnels, mines, and bridges were constructed prior to the early 1700s, when the first scientific foundations were laid for engineering. Engineers often must solve problems without even understanding the underlying theory. Certainly, engineers benefit from scientific theory, but sometimes the solution is required before the theory can catch up to the practice. For example, theorists are still trying to fully explain high-temperature superconductors while engineers are busy forming flexible wires out of these new materials that may be used in future generations of electrical devices.

2.3   THE NEED FOR ENGINEERING

Appendix C describes how humankind's needs have been met by engineering throughout history. As you prepare for a career in engineering, you should be aware of the problems you will face. Here, we look briefly at some of the challenges in our future.

2.3.1   Resource Stewardship and Utilization

The history of engineering can be viewed as "humans versus nature." Humans made progress when they overcame some of nature's terrors by redirecting rivers, paving land, felling trees, and mining the earth. In view of our large population (about 6 billion), we can claim victory.

The Trebuchet: An Engine of War

The trebuchet (pronounced *tray-boo-shay*) pictured below is an ancient "engine" of war. It consists of a long beam that rotates about a fixed fulcrum. One end of the beam has a cup or sling into which the projectile is placed. At the other end is a counterweight that, when released, causes the beam to rotate and throw the projectile into the air.

The trebuchet was invented in China about 2200 years ago and reached the Mediterranean about 1400 years ago. It could throw objects weighing up to 1 ton great distances; in fact, it was used even after the invention of the cannon, because its range was greater than that of early artillery. A modern trebuchet constructed in England could throw a 476-kg car (without engine) 80 meters using a 30,000-kg counterweight. Ancient machines threw stones, dead horses, and even diseased human corpses as a form of biological warfare.

As is often the case, practice preceded theory; trebuchets were constructed and used long before their theory was understood. Many modern concepts, such as force vectors and work (a force exerted over a distance), are thought to have been developed by engineers seeking to improve trebuchet performance. The trebuchet is an example of military necessity causing advances in scientific understanding, a process that is still occurring.

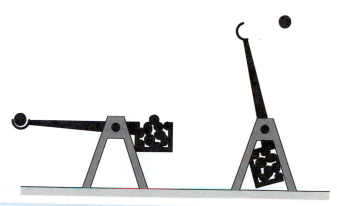

Adapted from P. E. Chevedden, L. Eigenbrod, V. Foley, and W. Soedel, "The Trebuchet," *Scientific American,* July 1995, pp. 66–71.

The rising wave of environmentalism results from our recognition that a fundamental change is now required. We can no longer be nature's adversary, but must become its caretaker. We have become so powerful, we literally can eliminate whole ecosystems either deliberately (e.g., by felling rain forests) or inadvertently (e.g., by releasing pollution into the water and air). Many scientists are also concerned that human activity may result in changing weather patterns due to the release of "greenhouse gases" such as carbon dioxide, methane, chlorofluorocarbons, and nitrogen oxides. Some chlorine-containing gases are implicated in the destruction of the ozone layer, which protects plants and animals from damaging ultraviolet light.

Although we humans have become extremely powerful, we still depend upon nature to provide the basics of life, such as food and oxygen. These basics do not come easily. NASA has spent millions of dollars to develop regenerative life support systems for use on the moon or Mars that allow people to live independently of earth's life support system. The research continues because the problem is so challenging.

"Sustainable development" is a recent economic philosophy that recognizes humans' right to live and improve their standard of living, while simultaneously protecting the environment. This philosophy attempts to reshape our economy to achieve sustainability. For example, basing our energy sources on fossil fuels is not sustainable. Eventually they will run out, or the pollution resulting from their use will make the planet uninhabitable. Sustainable development would require the use of renewable energy sources such as solar, wind, and biomass fuels, or "infinite" energy sources such as fission (with breeder reactors) or fusion. Resource conserving, recycling, and nonpolluting technologies are also essential to sustainable development.

In modern times, many resources are used once and then thrown away. This "one-pass" approach is increasingly unacceptable, because of the finite nature of our resources and because discarded resources cause pollution. Instead, engineers must develop a cyclical approach in which resources are reused. Some products are now designed to be dismantled when their useful life is completed. They are constructed of metals and polymers that can be reformed into new products.

All processes, including the cyclical processes developed by future engineers, are driven by energy. Because energy production expends resources and causes pollution, it is incumbent upon engineers to develop energy-efficient processes. Many of our current processes use energy inefficiently and can be greatly improved by future engineers.

Unavoidably, all processes produce waste. In the future, many engineers will be required to design processes that minimize wastes, produce wastes that can be converted to useful products, or convert the wastes to forms that can be safely stored.

2.3.2 Global Economy

During World War II, while much of the world economy was destroyed, the U.S. economy remained intact. For a few decades immediately following the war, the U.S. economy was very strong with high export levels. Foreign nations wanted our goods—not because they were of superior quality, but because there were few alternatives. In fact, the quality of many U.S. goods actually deteriorated due to sloppy manufacturing practices, adopted because our industry was not challenged by competition.

Today, the world economy is completely different. The economies of the world have long since recovered from the war. Many nations are capable of producing goods that are equal or superior to the quality of U.S. goods. After the war, a product labeled "Made in Japan" was assumed to be of poor quality; today, this label is an indication that the product is well made and affordable.

In a free market, consumers are able to buy products from all over the world. When they select products made in other countries, it represents a loss of jobs for the United States. American industry is meeting this challenge by instituting "quality" into the corporate culture. A company that is committed to quality must identify their customers, learn their requirements, and transform its manufacturing and management practices to create products that meet the customers' needs and expectations.

Because labor is generally less expensive overseas, many labor-intensive products cannot be economically manufactured in the United States using current technology. However, if engineers develop manufacturing methods that use machines to replace labor, then many of these products can be made in the United States.

Another way for the United States to compete is by developing high-technology products. A major U.S. competitive advantage is our very strong science base. We have a very healthy scientific enterprise in this nation. By translating the latest scientific research into consumer products, we can maintain a competitive edge.

2.4 THE TECHNOLOGY TEAM

Modern technical challenges are seldom met by the lone engineer. Technology development is a complex process involving the coordinated efforts of a technology team consisting of:

A Few Words on Diversity

To fully describe a person, the list of traits might include intellectual ability, personality, creativity, educational level, hobbies, hair color, skin color, body weight, height, age, physical strength, gender, religion, ethnic background, sexual orientation, nationality, language, parental upbringing, and so forth. The list is long, and there are so many variations within each trait that certainly every person is unique.

Because humanity is so diverse, you can be assured that the teammates on your technology team will be different from you. This diversity will be a source of either strength or weakness, depending upon how you respond to it.

Diversity is a source of strength when people with various backgrounds and abilities all work together on the technical problem. The benefits of diversity have long been recognized; hence the expression "two heads are better than one." This simple statement recognizes the fact that a single person may not have all the skills necessary to solve a complex problem, but collectively, the needed skills are there. Also, a diverse team has a useful variety of viewpoints. For example, although traditional automobile design teams have been strictly male, women have recently joined these teams. The female teammates have introduced a new perspective to automobile design, making the cars safer and more appealing to women, who constitute about 50% of the car-buying public.

Diversity is a source of weakness if teammates are so different that they cannot communicate, or they mistrust each other and cannot work together toward a common end. This potential weakness results from two common human tendencies: tribalism and overgeneralization. *Tribalism* refers to the fact that during most of human history, people have lived in tribes composed of similar members. When outsiders entered the tribal land, they were often treated with suspicion because they were potential enemies. *Overgeneralization* refers to the fact that in their attempt to understand the world, humans make generalizations from specific observations—but sometimes the generalizations go too far. For example, if Laura were watching a basketball game, she would observe that the team is composed primarily of tall people. After the game, if Laura were to meet Greg, who happens to be seven feet tall, she might assume that Greg plays basketball, when, in fact, he has no interest in the game. Tribalism and overgeneralization prevent people from dealing with each other as individuals; instead, perceived attributes of a group are automatically assigned to an individual. Not acknowledging the true character of a co-worker makes a working relationship impossible.

To gain strength from diversity and avoid potential pitfalls, it is important that the technology team share a common set of core values that allow it to work together. Some sample core values are shown below:

- Teammates are rewarded on the basis of hard work, not politics.
- Teammates are treated with respect.
- Teammates are treated as unique persons with their own skills, talents, abilities, and perspectives.

Adopting these core values, and others, will allow the team to function in harmony and gain strength from diversity.

- *Scientists,* who study nature in order to advance human knowledge. Although some scientists work in industry on practical problems, others have successful careers publishing results that may not have immediate practical applications. Typical degree requirement: BS, MS, PhD.
- *Engineers,* who apply their knowledge of science, mathematics, and economics to develop useful devices, structures, and processes. Typical degree requirement: BS, MS, PhD.
- *Technologists,* who apply science and mathematics to well-defined problems that generally do not require the depth of knowledge possessed by engineers and scientists. Typical degree requirement: BS.
- *Technicians,* who work closely with engineers and scientists to accomplish specific tasks such as drafting, laboratory procedures, and model building. Typical degree requirement: two-year associate's degree.
- *Artisans,* who have the manual skills (welding, machining, carpentry) to construct devices specified by scientists, engineers, technologists, and technicians. Typical degree requirement: high school diploma plus experience.

Elijah McCoy: Mechanical Engineer and Inventor

Elijah McCoy was born in the early 1840s in Colchester, Ontario, Canada. His parents were former slaves who escaped from Kentucky via the Underground Railroad, a network of individuals who helped slaves reach freedom.

At that time, educational opportunities for blacks were limited, so at age 15, McCoy's parents sent him to study in Scotland, where he achieved the title "master mechanic and engineer." He returned to North America and settled in Detroit, Michigan. During the 1860s, it was difficult for blacks to obtain jobs in the professions, so his first job was a fireman/oilman on the Michigan Central Railroad. As a fireman, he shoveled coal into the firebox. As an oilman, he lubricated the machinery, which had to be stopped for that purpose, causing delays and reducing efficiency. This experience inspired his first patent (U.S. Patent 129,843, issued July 12, 1872), for a device that lubricated machinery while in motion. This lubricating device was so superior to the competition that some engineers would ask if machinery was equipped with *the real McCoy,* a popular American expression meaning *the real thing.* Interestingly, this expression originated in an 1856 advertising slogan *the real MacKay,* used to promote a Scottish brand of whiskey.

During his life, McCoy developed 57 patents. They were issued in the United States, Great Britain, Canada, France, Germany, Austria, and Russia. Among them were an ironing board and a lawn sprinkler.

In 1920, he established the Elijah McCoy Manufacturing Company to manufacture and sell his numerous inventions. He died nine years later in 1929. To honor his achievements as an inventor, he was inducted into the National Inventors Hall of Fame in 2001.

Adapted from the following websites: web.mit.edu/invent/iow/mccoy.html, www.blackinventor.com/pages/elijahmccoy.html, www.inventions.com/culture/african/mccoy.html

Successful teamwork results in accomplishments larger than can be produced by individual team members. There is a magic when a team coalesces and each member builds off of the ideas and enthusiasm of teammates. For this magic to occur and to produce output that surpasses individual efforts, several characteristics must be present:

- Mutual respect for the ideas of fellow team members.
- The ability of team members to transmit and receive the ideas of the team.
- The ability to lay aside criticism of an idea during early formulation of solutions to a problem.
- The ability to build on initial or weakly formed ideas.
- The skill to accurately criticize a proposed solution and analyze for both strengths and weaknesses.
- The patience to try again when an idea fails or a solution is incomplete.

2.5 ENGINEERING DISCIPLINES AND RELATED FIELDS

At this point in your engineering career, you may not have selected a major. Does your future lie in mechanical engineering, chemical engineering, electrical engineering, or other engineering fields? Once you have made your selection, you will have decided upon your engineering *discipline.* To help in this decision, we briefly describe the major engineering disciplines and some related fields.

Figure 2.1 shows when the major engineering disciplines were born. Nearly all disciplines are thought to have evolved from civil engineering. Note that all engineering disciplines require extensive knowledge of physics, whereas chemical and materials engineering require extensive knowledge of physics and chemistry. Some recent disciplines (biochemical and biomedical) require extensive knowledge of physics, chemistry, and biology.

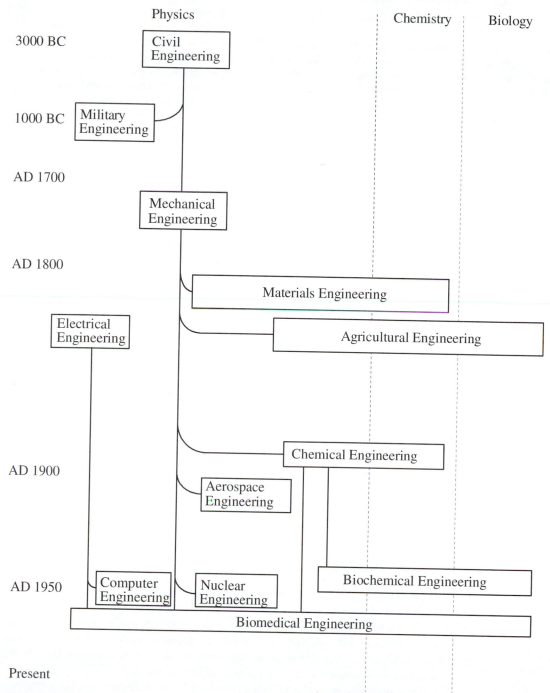

FIGURE 2.1
Birth of engineering disciplines (birth dates are approximate).

Josephine Garis Cochrane: Inventor of the Dishwasher

In 1839, Josephine Garis was born into an industrious family. Her father, John Garis, was a civil engineer who supervised mills along the Ohio River and drained swamps to develop Chicago during the 1850s. Her great-grandfather, John Fitch, built a steamboat of his own design that served Philadelphia in 1786.

In 1853, at age 19, Josephine Garis married William Cochran, a handsome 27-year-old man who became wealthy in the dry goods business. Josephine was an independent woman—although she took her husband's name, she insisted on ending it with an *e*.

The young socialite couple was popular and had many friends, whom they entertained frequently with elaborate dinner parties using family heirloom china. The servants who washed the china were careless and broke too many plates, so Josephine decided to wash and dry the dishes herself. She soon concluded that this activity wasted her precious time, so she resolved to design a machine that would wash the dishes for her. Within a half hour, she decided that the machine should hold the dishes in a rack and high-pressure water would scrub them clean.

Shortly thereafter, in 1883, Josephine's husband died. Although he was a wealthy man, he had spent more than he earned, leaving her destitute. Nonetheless, with the help of mechanic George Butters, she built the first dishwasher in the shed behind her home, a site now marked with a historical marker. Powered by a hand pump, it cleaned dishes using streams of soapy water. Friends and neighbors came to see the contraption. They were delighted and encouraged her to pursue it further. On December 28, 1886, Mrs. Cochrane received her first patent on the dishwasher.

Because it was expensive, she decided to market her dishwasher to institutions, rather than homes. The Palmer House, a famous Chicago hotel, was her first customer. Her dishwasher could wash and dry 240 dishes within 2 minutes.

Because she had no capital, she hired a contractor to build the units. Relations were strained; the contractor would often ignore her ideas because she had no formal mechanical training and because she was a woman. Further, the contractor took most of the profits, even though she had the brains, patents, entrepreneurial talent, and sales orders. She was not able to raise capital from investors because they refused to invest in a company headed by a woman.

In 1893, nine of her Garis-Cochran dishwashers cleaned dirty dishes at the World's Columbian Exposition, a large fair held in Chicago. Judges awarded her machine the highest prize, stating it had the "best mechanical construction, durability and adaptation to its line of work." The resulting publicity generated more orders. By 1898, Mrs. Cochrane had saved enough money to open her own manufacturing facility and no longer depend on a contractor to build her machines. George Butters became the foreman and oversaw the three employees. Finally, she could work with people who respected her and did not challenge her ideas. Her dishwashers were acclaimed by hotels and other institutions because they saved labor, reduced breakage, and sanitized the dishes. Josephine had succeeded and lived to see her business thrive.

After her death in 1913, the company continued to manufacture dishwashers of her design. In 1926, the company was purchased by Hobart, a manufacturer of well-engineered appliances. Hobart changed the name of the dishwasher subsidiary to KitchenAid, which finally introduced a home dishwasher in the 1940s. Later, KitchenAid was acquired by Whirlpool, a major home appliance manufacturer.

Adapted from J. M. Fenster, "The Woman Who Invented the Dishwasher," *American Heritage of Invention & Technology,* vol. 15, no. 2, pp. 54–61, Fall 1999.

2.5.1 Civil Engineering

Civil engineering is generally considered the oldest engineering discipline—its works trace back to the Egyptian pyramids and before. Many of the skills possessed by civil engineers (e.g., building walls, bridges, roads) are extremely useful in warfare, so these engineers worked on both military and civilian projects. To distinguish those engineers who work on civilian projects from those who work on military projects, the British engineer John Smeaton coined the term *civil engineer* in about 1750.

Civil engineers are responsible for constructing large-scale projects. Civil engineers must consider many factors in their designs, such as construction costs, schedules, government regulations, environmental impacts, and potential hazards (e.g., hurricanes, earthquakes). Major specialties are listed below.

Ancient Egypt: From Engineer to God

Egyptian civilization ascended from the Late Stone Age, around 3400 B.C., with vigorous advancements in several engineering fields. While we can still see the spectacular construction feats of the Pyramid Age (3000–2500 B.C.), the ancient Egyptians also pioneered other engineering fields. As hydraulic engineers, they manipulated the Nile River for agricultural and commercial purposes; as chemical engineers, they produced dyes, cement, glass, beer, and wine; as mining engineers, they extracted copper from the Sinai Peninsula for use in the bronze tools that built the pyramids.

One of the key players of this period was Imhotep, known today as "The Father of Stone Masonry Construction." Imhotep served the pharaoh Zoser as chief priest, magician, physician, and head engineer. Most archaeologists credit Imhotep with designing and building the first pyramid, a stepped tomb for Zoser at Sakkara, around 2980 B.C. This pyramid consists of six stages, each 30 feet high, built from local limestone, and hewn with copper chisels. While only 200 feet high (the height of an 18-story building), this unique structure served as a prototype for the Great Pyramid at Giza, constructed 70 years later, which covers four city blocks in area and originally stood 480 feet high.

Imhotep acquired an extensive reputation as a sage, and in later centuries was recognized as the Egyptian god of healing. Although Egyptian civilization saw great engineering progress during the Pyramid Age, 2000 years of stagnation and decline followed.

Hypostyle Hall of Karnak Temple in Luxor, Egypt.

Photo © Richard Passmore/Getty Images.

Courtesy of Seth Adelson

- *Construction engineers* use technical and management skills to build projects on time and within budget. Their abilities include construction methods, equipment, financing, planning, and management.
- *Environmental engineers* protect our planet from wastes generated in our industrial society. Environmental engineers use physical, chemical, and biological processes to provide safe drinking water, clean up contaminated sites, prevent air pollution, treat wastewater, and manage solid wastes.
- *Geotechnical engineers* are experts in soil mechanics. Their skills are essential when constructing tunnels, foundations, pipelines, highways, earth dams, levees, and embankments.

- *Structural engineers* analyze and design structures to ensure that they are safe not only under their own load, but also when subjected to dynamic loads, such as hurricanes, earthquakes, blizzards, and floods. These engineers design buildings, offshore platforms, space stations, amusement park rides, and bridges. They must be familiar with the properties of building materials (e.g., steel, concrete, aluminum, timber).
- *Transportation engineers* are involved with systems that move people and things by land, air, and sea. They design, construct, and maintain highways, railroads, airports, and seaports.
- *Urban planners* develop communities by planning streets, recreational parks, industrial parks, and residential areas. They must coordinate with community authorities as well as citizens. To ensure success, they must have good people and technical skills.
- *Water resource engineers* design systems that prevent floods, distribute water, purify drinking water, treat wastewater, protect beaches, and manage rivers. Their projects include hydroelectric plants, canals, dams, pipelines, pumping stations, locks, and seaports.

Because of their close association with civil projects, many civil engineers work in government. Others work with engineering construction firms that build the large projects. Some have their own consulting businesses.

2.5.2 Mechanical Engineering

Mechanical engineering was practiced concurrently with civil engineering because many of the devices needed to construct great civil engineering projects were mechanical in nature. During the Industrial Revolution (1750–1850), wonderful machines were developed: steam engines, mechanical looms, sewing machines, and more. Here we saw the birth of mechanical engineering as a discipline distinct from civil engineering.

Because machines are involved in almost all aspects of life, mechanical engineering is among the most versatile engineering disciplines. Specialties include the following.

- *Energy production* involves creating useful energy (e.g., shaft rotation, electricity) from available resources (e.g., fuels, wind, sun). This specialty involves the design, construction, and testing of engines (e.g., internal combustion engines, steam engines, gas turbines), wind turbines, and solar systems.
- *Machines* are designed, constructed, and tested by mechanical engineers to perform many functions. For example, manufacturing plants require specialized machines to transform raw materials into finished goods that are packaged for sale. Mines use specialized machines to recover ores from the earth. The military requires machines to protect our nation. Consumers use many machines (e.g., dishwashers, clothes washers, hair dryers), all of which are designed by mechanical engineers.
- *Manufacturing* involves the use of machine tools (e.g., lathes, mills, shapers, grinders) to manufacture components that are assembled into finished products. Mechanical engineers in this specialty must be familiar with many processes, such as metal cutting, grinding, welding, forging, injection molding, and ceramic fabrication. Computers control many machine tools so that components can be made accurately and repeatably.
- *Transportation* involves the design, construction, and testing of transportation systems, such as automobiles, planes, rockets, trains, buses, and ships.
- *Heating, ventilation, air conditioning, and refrigeration* engineering is a specialty that involves the design, construction, and testing of systems that heat and cool buildings, as well as refrigerators and freezers.

- *Heat transfer* is a specialty that focuses on the design, construction, and testing of heat exchangers that transfer heat from a hot fluid to a cold fluid.
- *Fluid flow* is a specialty that involves the flow of fluids inside objects (e.g., pipes, stirred tanks) or outside of objects (e.g., planes, automobiles). Sophisticated computational fluid dynamic (CFD) software predicts fluid flow, which can be verified experimentally in wind tunnels.

Because the field is so broad, the above list is just a sampling of the many opportunities available to mechanical engineers.

2.5.3 Electrical Engineering

Soon after physicists began to understand electricity, the electrical engineering profession was born. Early electrical engineers were involved with installing electric grids (primarily to light homes) and telegraph system for communications. Later, they began developing more sophisticated systems, such as telephones and radios.

The impact of electrical engineering is dramatic; modern life is largely characterized by electronic equipment. Daily, we rely on many electronic devices: televisions, telephones, computers, calculators, and so on. In the future, the number and variety of these devices can only increase.

Electronic equipment can be *analog* (meaning the voltages and currents in the device are *continuous* values) or *digital* (meaning only *discrete* voltages and currents can be attained by the device). Because analog equipment is more susceptible to noise and interference than digital equipment, many electrical engineers specialize in digital circuits.

Electricity serves two main functions in society: the transmission of power and of information.

Specialties in electrical engineering include those listed below.

- *Power generation and transmission* involves generating electricity at centralized power plants or distributed sites (e.g., wind, solar), which is then transmitted to consumers through an electrical grid. Electrical engineers design and operate the electrical system, which includes generators, transformers, power lines, switches, and circuit breakers.
- *Motors* consume electricity to produce motion. Most motors convert electricity into rotary motion; however, a few produce linear motion. Modern motors can be controlled with power electronics that allows the speed to vary. The introduction of hybrid electric automobiles is opening more opportunities in this field.
- *Electronics* involves components (e.g., capacitors, inductors, switches, diodes, transistors, integrated circuits) that modify the flow of electrons. Electrical engineers assemble these components into circuits used in devices, such as security systems, televisions, and cell phones.
- *Computers* use digital electronic circuits that control the flow of electrons for computational purposes. Electrical engineers who specialize in computers work with integrated circuits (e.g., motherboards), input devices (e.g., keyboards), memory devices (e.g., hard disks), pointing devices (e.g., mouse), and display devices (e.g., flat-panel displays).
- *Communications systems* involve networks that transmit information through the air (via radio or microwaves) or through fiber-optic cables (via light).

- *Instrumentation and measurement* involves sensors that measure various physical phenomena, such as temperature, rotation rate, light intensity, pressure, and vibration. Calibrated instruments translate the sensor signals into meaningful numbers.
- *Automatic controls* involves changing an input to obtain a desired output. For example, suppose we desire to regulate the temperature of the room. The heating rate would be adjusted to obtain the desired room temperature. Some control strategies will allow the room temperature to stay within a narrow band, whereas other strategies allow the room temperature to vary widely. This specialty tends to be highly mathematical.

2.5.4 Chemical Engineering

By 1880, the chemical industry was becoming important in the U.S. economy. At that time, the chemical industry hired two types of technical persons: mechanical engineers and industrial chemists. The chemical engineer combined the skills of these two persons into one. In 1888, the first chemical engineering degree was offered at the Massachusetts Institute of Technology.

Chemical engineering is characterized by a concept called *unit operations*. A unit operation is an individual piece of process equipment (e.g., chemical reactor, heat exchanger, pump, compressor, distillation column). Just as electrical engineers assemble complex circuits from component parts (resistors, capacitors, inductors, and batteries), chemical engineers assemble chemical plants by combining unit operations together.

As with all engineers, chemical engineers must understand physics. In addition, chemical engineers must understand chemistry. Increasingly, they must understand biology as well. Because of their strong foundation in the basic sciences, chemical engineers are extremely versatile.

A few chemical engineering specialties are listed below.

- *Petrochemical engineers* convert crude oil and natural gas into finished products, such as gasoline, diesel fuel, heating oil, industrial chemicals, solvents, and polymers.
- *Pharmaceutical engineers* produce high-value pharmaceuticals from raw materials. Some pharmaceuticals (e.g., aspirin) are made using traditional chemistry, but increasingly, pharmaceuticals are made using biological processes that involve enzymes (e.g., steroids) or genetically engineered microorganisms (e.g., human insulin).
- *Fine chemical engineers* produce high-value, small-volume chemicals (e.g., dyes, flavors, fragrances).
- *Food engineers* produce finished foods from raw ingredients. For example, chemical engineers are involved in the various steps required to convert a raw potato into potato chips.
- *Safety engineers* ensure that chemical processes operate safely. This specialty involves identifying ways that chemical plants can fail and installing systems that prevent failure or minimize damage should a failure occur.
- *Environmental engineers* reduce pollution emitted from chemical plants and refineries.
- *Design engineers* design chemical plants and refineries using sophisticated computer software. They assemble unit operations into processes and evaluate their economic viability.

2.5.5 Industrial Engineering

In the late 1800s, industries began to use "scientific management" techniques to improve efficiency. Early pioneers in this field did time-motion studies on workers to reduce the amount of labor required to produce a product. Today, industrial engineers develop, design, install, and operate integrated systems of people, machinery, and information to produce either goods or services. Industrial engineers bridge engineering and management.

Industrial engineers develop processes and systems that improve quality and productivity with the objective of eliminating waste (e.g., time, money, materials, energy). Industrial engineers are systems integrators who focus on the big picture rather than minutiae. It is a flexible profession that is valued by many industries.

A few industrial engineering specialties follow.

- *Methods engineers* perform time-motion studies to minimize wasted effort by workers. They ensure that workers have the appropriate tools, that materials are readily accessible, and that the working environment is safe and comfortable.
- *Plant layout engineers* design efficient manufacturing facilities that make the best use of space and minimize the distance that components must move during assembly. These engineers are often involved in the design and operation of assembly lines.
- *Cost engineers* study a manufacturing process and determine ways to minimize costs, thereby increasing profitability.
- *Quality engineers* use statistical methods to study variations in the way that processes operate. Their objective is to minimize variability so that products have the same quality consistently.
- *Human factors engineers* ensure that people interact with machines and processes in an optimal manner. They consider not only physiological factors, but psychological factors as well.
- *Operations research engineers* use mathematics to optimize processes, such as those that involve inventory, schedules, and queues. Maintaining an inventory is very expensive; however, it is desirable to have some parts on hand because an operation can be shut down while waiting for an out-of-stock part. Knowing what and how many parts to stock requires study and optimization. Some modern plants use just-in-time manufacturing, in which almost no parts are stored in inventory. In some cases, parts are delivered daily. Schedules make optimal use of assets by establishing proper sequences of operation, such as plane flights. A queue is a collection of items in which only the earliest added item may be accessed. A classic example is waiting in line at a bank. Do you have a separate line for each teller, or do you have a single line that supplies all the tellers with customers? Operations research engineers allow many people-oriented industries (e.g., hospitals, banks, hotels, fast-food restaurants) to operate more efficiently.

2.5.6 Aerospace Engineering

With their first successful flight in 1903, the Wright brothers ushered in *aeronautical engineering,* the field that focuses on vehicles that fly in the atmosphere. In 1957, the launch of Sputnik 1 ushered in *astronautical engineering,* the field that focuses on space flight. Aerospace engineering encompasses both of these disciplines. Aerospace engineers may be viewed as mechanical engineering specialists who design, develop, manufacture, and test aircraft, spacecraft, and missiles. A few aerospace specialties follow.

- *Aerodynamics engineers* study fluid flow over solid surfaces. They determine the optimal shape of a vehicle to minimize drag and maximize controllability while it travels through a fluid. Most commonly, the vehicle is a plane or missile and the fluid is air; however, aerodynamic principles may be applied to vehicles that travel through water, such as submarines. Also, aerodynamics can be used to design blades for windmills or to study wind forces on buildings. In the design phase of a project, aerodynamics engineers use sophisticated computational fluid dynamics (CFD) software to predict the drag on a vehicle. Also, they can test their designs in a wind tunnel, where scaling laws are used to predict the performance of full-size vehicles from measurements made on small models.
- *Propulsion engineers* design, build, and test vehicle propulsion systems, such a piston-driven propellers, jet engines, and rocket engines. These propulsion systems must have a high power density. Ideally, they should use fuel efficiently to minimize the amount of fuel and maximize the payload.
- *Structural engineers* must determine the forces on the vehicle from the payload and fuel, as well as from turbulence and landing. They design the airframe to handle the forces with the minimum weight of material.
- *Control engineers* design methods to control the vehicle using winged surfaces or thrust vectoring. In some cases, the objective is optimal stability (e.g., passenger plane), and in other case, the objective is optimal maneuverability (e.g., military fighter).

2.5.7 Materials Engineering

Materials engineers are concerned with obtaining the materials required by modern society. Materials engineers may be further classified as:

- *Geological engineers,* who study rocks, soils, and geological formations to find valuable ores and petroleum reserves.
- *Mining engineers,* who extract ores such as coal, iron, and tin.
- *Petroleum engineers,* who find, produce, and transport oil and natural gas.
- *Ceramic engineers,* who produce ceramic (i.e., nonmetallic mineral) products.
- *Plastics engineers,* who produce plastic products.
- *Metallurgical engineers,* who produce metal products from ores or create metal alloys with superior properties.
- *Materials science engineers,* who study the fundamental science behind the properties (e.g., strength, corrosion resistance, conductivity) of materials.

2.5.8 Agricultural Engineering

Agricultural engineers help farmers efficiently produce food and fiber. This discipline was born with the McCormick reaper. Since then, agricultural engineers have developed many other farm implements (tractors, plows, choppers, etc.) to reduce farm labor requirements. Modern agricultural engineers apply knowledge of mechanics, hydrology, computers, electronics, chemistry, and biology to solve agricultural problems. Agricultural engineers may specialize in food and biochemical engineering; water and environmental quality; machine and energy systems; and food, feed, and fiber processing.

2.5.9 Nuclear Engineering

Nuclear engineers design systems that employ nuclear energy, such as nuclear power plants, nuclear ships (e.g., submarines and aircraft carriers), and nuclear spacecraft. Some nuclear engineers are involved with nuclear medicine; others are working on the design of fusion reactors that potentially will generate limitless energy with minimal environmental damage.

2.5.10 Architectural Engineering

Architectural engineers combine the engineer's knowledge of structures, materials, and acoustics with the architect's knowledge of building esthetics and functionality.

2.5.11 Biomedical Engineering

Biomedical engineers combine traditional engineering fields (mechanical, electrical, chemical, industrial) with medicine and human physiology. They develop prosthetic devices (e.g., artificial limbs), artificial kidneys, pacemakers, and artificial hearts. Recent developments will enable some deaf people to hear and some blind people to see. Biomedical engineers can work in hospitals as clinical engineers, in medical centers as medical researchers, in medical industries designing clinical devices, in the FDA evaluating medical devices, or as physicians providing health care.

2.5.12 Computer Science and Engineering

Computer science and engineering evolved from electrical engineering. Computer scientists understand both computer software and hardware, but they emphasize software. In contrast, computer engineers understand both computer software and hardware but emphasize hardware. Computer scientists and engineers design and build computers ranging from supercomputers to personal computers, network computers together, write operating system software that regulates computer functions, or write applications software such as word processors and spreadsheets. Given the increasingly important role of computers in modern society, computer science and engineering are rapidly growing professions.

2.5.13 Engineering Technology

Engineering technologists bridge the gap between engineers and technicians. Engineering technologists typically receive a 4-year BS degree and share many courses with their engineering cousins. Their course work evenly emphasizes both theory and hands-on applications, whereas the engineering disciplines described above primarily emphasize theory with less emphasis on hands-on applications. Engineering technologists can acquire specialties such as general electronics, computers, and mechanics. With their skills, engineering technologists perform such functions as designing and building electronic circuits, repairing faulty circuits, maintaining computers, and programming numerically controlled machine shop equipment.

2.5.14 Engineering Technicians

Engineering technicians typically receive a 2-year associate's degree. Their education primarily emphasizes hands-on applications with less emphasis on theory. They are involved in product design, testing, troubleshooting, and manufacturing. Their specialties include the following: electronics, drafting, automated manufacturing, robotics, and semiconductor manufacturing.

2.5.15 Artisans

Artisans often receive no formal schooling beyond high school. Typically, they learn their skills by apprenticing with experienced artisans who show them the "tricks of the trade." Artisans have a variety of manual skills such as machining, welding, carpentry, and equipment operation. Artisans are generally responsible for transforming engineering ideas into reality; therefore, engineers often must work closely with them. Wise engineers highly value the opinions of artisans, because artisans frequently have many years of practical experience.

2.5.16 Engineering Employment Statistics

Table 2.1 shows the number of engineers employed in the United States. Approximately 1.0 percent of all employees are engineers.

2.6 ENGINEERING FUNCTIONS

Regardless of their discipline, engineers can be classified by the functions they perform:

- *Research engineers* search for new knowledge to solve difficult problems that do not have readily apparent solutions. They require the greatest training, generally an MS or PhD.
- *Development engineers* apply existing and new knowledge to develop prototypes of new devices, structures, and processes.
- *Design engineers* apply the results of research and development engineers to produce detailed designs of devices, structures, and processes that will be used by the public.
- *Production engineers* are concerned with specifying production schedules, determining raw materials availability, and optimizing assembly lines to mass produce the devices conceived by design engineers.
- *Testing engineers* perform tests on engineered products to determine their reliability and suitability for particular applications.
- *Construction engineers* build large structures.
- *Operations engineers* run and maintain production facilities such as factories and chemical plants.
- *Sales engineers* have the technical background required to sell technical products.
- *Managing engineers* are needed in industry to coordinate the activities of the technology team.
- *Consulting engineers* are specialists who are called upon by companies to supplement their in-house engineering talent.
- *Teaching engineers* educate other engineers in the fundamentals of each engineering discipline.

TABLE 2.1
Number of engineers and other professions in the United States (2004)

Engineers	1,450,100
Civil	237,000
Mechanical	226,000
Industrial	177,000
Electrical	156,000
Electronics, except computer	143,000
Computer hardware	77,000
Aerospace	76,000
Environmental	49,000
Chemical	31,000
Health and safety	27,000
Materials	21,000
Nuclear	17,000
Petroleum	16,000
Biomedical	9,700
Marine and naval architecture	6,800
Mining and geological	5,200
Agricultural	3,400
Other	172,000

Other Professionals	1,811,000
Lawyers	735,000
Physicians	567,000
Pharmacists	230,000
Dentists	150,000
Architects	129,000

Scientists	183,000
Chemists and material scientists	90,000
Biologists	77,000
Physicists and astronomers	16,000

Teachers	5,430,000
Preschool	431,000
Kindergarten	171,000
Elementary	1,500,000
Middle	628,000
Secondary	1,100,000
Post secondary	1,600,000

Total all workers	139,252,000
Total U.S. population	296,000,000

Sources: http://www.bls.gov/oco/home.htm
        http://quickfacts.census.gov/qfd/states/00000.html
        http://www.census.gov/compendia/statab/tables/06s0604.xls

To illustrate the roles of engineering disciplines and functions, consider all the steps required to produce a new battery suitable for automotive propulsion. (The probable engineering discipline is in parentheses and the engineering function is in italics.) A *research engineer* (chemical engineer) performs fundamental laboratory studies on new materials that are possible candidates for a rechargeable battery that is lightweight but stores much energy. The *development engineer* (chemical or electrical engineer) reviews the results of the research engineer and selects a few candidates for further development. She constructs some battery prototypes and tests them for such properties as maximum number of recharge cycles, voltage output at various temperatures, effect of discharge rate on battery life, and corrosion. If the development engineer lacks expertise in corrosion, the company would temporarily hire a *consulting engineer* (chemical, mechanical, or materials engineer) to solve a corrosion problem. When the development engineer has finally amassed sufficient information, the *design engineer* (mechanical engineer) designs each battery model that will be produced by the company. He must specify the exact composition and dimension of each component and how each component will be manufactured. A *construction engineer* (civil engineer) erects the building in which the batteries will be manufactured and a *production engineer* (industrial engineer) designs the production line (e.g., machine tools, assembly areas) to mass produce the new battery. *Operations engineers* (mechanical or industrial engineers) operate the production line and ensure that it is properly maintained. Once the production line is operating, *testing engineers* (industrial or electrical engineers) randomly select batteries and test them to ensure that they meet company specifications. *Sales engineers* (electrical or mechanical engineers) meet with automotive companies to explain the advantages of their company's battery and answer technical questions. *Managing engineers* (any discipline) make decisions about financing plant expansions, product pricing, hiring new personnel, and setting company goals. All of these engineers were trained by *teaching engineers* (many disciplines) in college.

In this example, the engineering disciplines that satisfy each function are unique to the project. Other projects would require the coordinated efforts of other engineering disciplines. Also, the disciplines selected for this project are an idealization. A company might not have the ideal mix of engineers required by a project and would expect its existing engineering staff to adapt to the needs of the project. After many years, engineers become cross trained in other disciplines, so it becomes difficult to classify them by the disciplines they studied in college. An engineer who wishes to stay employed must be adaptable, which means being well acquainted with the fundamentals of other engineering disciplines.

2.7   HOW MUCH FORMAL EDUCATION IS RIGHT FOR YOU?

Knowledge is expanding at an exponential rate. It is impossible to fully grasp engineering in a 4-year BS degree. Although you will continue learning on the job, your experience there will tend to be narrowly focused on the needs of the company.

As you proceed through your engineering studies, you should ask yourself, How much more formal education do I need? The answer depends upon your ultimate career objectives. Many of the job functions described above can be performed adequately with a BS degree. However, others—like the research engineer and the development engineer—generally require an MS or a PhD. These individuals are engaged in the early stages of product development. More education is required because they must solve more challenging technical problems.

If you think that you would enjoy the technical challenges met by advanced-degree engineers, do not let the educational costs dissuade you. Most graduate schools provide financial assistance to their students in the form of a stipend. Although the stipend does not equal the pay received in industry, it is usually enough to live a comfortable life. Because people with advanced degrees generally earn higher salaries (Figure 2.2), the short-term financial loss may eventually be recouped. Financial gain should not be your primary motivation for obtaining an advanced degree, however. You should consider it only if you would enjoy a job with greater technical challenges.

Some BS engineering students decide to continue formal education in other fields such as law, medicine, or business. The engineering curriculum provides an excellent background for these other fields because it develops excellent discipline, work habits, and thinking skills.

2.8   THE ENGINEER AS A PROFESSIONAL

Historically, a professional was simply a person who professed to be "duly qualified" in a given area. Often, these professionals professed adherence to the monastic vows of a religious order. So, being a professional meant not only mastering a body of knowledge, but also abiding by proper standards of conduct.

In the modern world, our concept of a professional has become more formalized. We consider a **professional** to have the following traits:

- *Extensive intellectual training*—all professions require many years of schooling, at the undergraduate or post-graduate level.
- *Pass qualifying exam*—professionals must demonstrate that they master a common body of knowledge.
- *Vital skills*—the skills of professionals are vital to the proper functioning of society.

FIGURE 2.2

Median salaries for engineers with different levels of education.

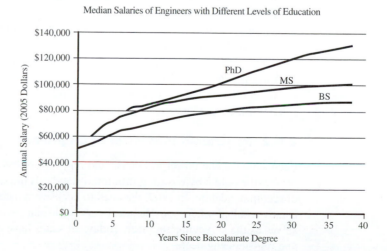

Median Salaries of Engineers with Different Levels of Education

- *Monopoly*—society gives professionals a monopoly to practice in their respective fields.
- *Autonomy*—society entrusts professionals to be self-regulated.
- *Code of ethics*—the behavior of professionals is regulated by self-imposed codes.

Engineering, architecture, medicine, law, dentistry, and pharmacy are examples of professions; they are some of the most prestigious occupations in our society.

2.8.1 Engineering Education

Since 1933, engineering education has been accredited by the Accrediting Board for Engineering and Technology (ABET). The primary purpose of accreditation is to ensure that graduates from engineering programs are adequately prepared to practice engineering. Although schools can offer nonaccredited engineering programs, graduates from these schools may have difficulty finding employment.

When an engineering program is evaluated by ABET, the evaluation team assesses the quality of the students, faculty, facilities, and curriculum. The curriculum must include (1) general education courses, (2) 1 year of college-level mathematics and basic sciences, and (3) $1\frac{1}{2}$ years of engineering science and design. The curriculum must culminate with a major design experience that employs realistic constraints from the following list: economic, environmental, sustainability, manufacturability, ethical, health and safety, social, and political.

Rather than prescribing a list of courses, ABET allows each engineering department to design its own curriculum that allows students to meet specified goals. During the evaluation of an engineering program, ABET determines if the graduates have the following skills:

a. An ability to apply knowledge of mathematics, science, and engineering.
b. An ability to design and conduct experiments, as well as to analyze and interpret data.
c. An ability to design a system, component, or process to meet desired needs.
d. An ability to function on multidiscipline teams.
e. An ability to identify, formulate, and solve engineering problems.
f. An understanding of professional and ethical responsibility.
g. An ability to communicate effectively.
h. The broad education necessary to understand the impact of engineering solutions in a global and societal context.
i. A recognition of the need for, and an ability to engage in, life-long learning.
j. A knowledge of contemporary issues.
k. An ability to use techniques, skills, and modern engineering tools necessary for engineering practice.

2.8.2 Registered Professional Engineer

Each state has the power to license and register professional engineers. The purpose is to protect the public by ensuring minimum standards through testing, experience, and letters of recommendation. In 1907, the need to license engineers was made evident in Wyoming. Chaos resulted when homesteaders surveyed their own water rights and declared themselves as the surveying engineer. Today, all states have an engineering board that licenses and registers engineers.

The Roman Republic and Empire: Paving the World

Over a period of 800 years, the city-state of Rome grew from a crowded Latin settlement to the nerve center of an empire stretching from present-day Scotland to Israel. To maintain the stability of their vast realm, the Romans implemented many public works, using contemporary technology to supply water, remove sewage, allow transportation, traverse rivers, and provide entertainment.

Scientific achievement under Roman rule was minimal; the Romans were not interested in theory. Applying simple principles with plenty of cheap materials and slave labor gave satisfactory results. For example, early Roman builders used the semicircular arch, an architectural concept developed by the Etruscans (a non-Indo-European people from northern Italy), to construct the magnificent aqueducts that supplied Rome with water. Although many earlier peoples had used concrete, Roman engineers manufactured an improved mixture, yielding a building material as hard and as waterproof as natural rock. With this improved concrete, they built well-planned cities with apartment buildings, or *insulae* ("islands"), that rose five stories high and provided central heating.

The Roman network of roads, beginning with the famed Via Appia in central Italy and then expanding outward into the Empire, was originally intended for military use. To defend its borders and continue its expansion, the Empire required rapid transportation of soldiers over a hard surface with sure footing. Roman engineers built their roads to last; they used simple instruments with plumb bobs to keep the surfaces level, and they often laid down four or five layers, 4 feet thick and 20 feet wide.

© Royalty Free/Corbis

Roman aqueduct in Segovia, Spain.

Courtesy of Seth Adelson.

An engineer does not need a license to practice engineering, but those who do have licenses have more career opportunities. Many industrial and government positions can only be filled by licensed engineers.

Although each state has its own licensing regulations, the procedure is generally as follows:

1. Obtain a degree from an institution recognized by the state engineering board. This requirement is automatically satisfied if the institution is accredited by ABET.
2. Successfully complete the Fundamentals of Engineering examination. This is an 8-hour exam on discipline specifics, as well as fundamentals in chemistry, mathematics, structures, electronics, economics, and other subjects. The title "Engineer in Training" (EIT) is given to engineering graduates who pass the exam.
3. Work 4 years as an engineer.
4. Obtain letters of recommendation.
5. Successfully complete the Principles and Practice examination, which is another 8-hour exam on the engineer's discipline.

Both exams are prepared by the National Council of Examiners for Engineering and Surveying (NCEES) and are offered throughout the country at about the same time. If you wish to become a registered professional engineer, you should plan to take the Fundamentals exam during your last semester of college, when the knowledge is still fresh in your mind.

2.8.3 Professional Societies

Most professions have professional societies. The American Medical Association (for physicians) and the American Dental Association (for dentists) serve the interests of those professions. Similarly, we engineers have professional societies that serve our interests. The first engineering professional society was the Institute of Civil Engineers, founded in Britain in 1818. The first American professional society was the American Society of Civil Engineers, founded in 1852. Since then, many other professional societies have been founded (Table 2.2).

The primary function of professional societies is to exchange information between members. This is accomplished in such ways as publishing technical journals, holding technical conferences, maintaining technical libraries, teaching continuing education courses, and providing employment statistics (salaries, fringe benefits) so members can assess their compensation. Some professional societies will assist members to find jobs or advise government in technical matters related to their profession.

As a student, you are highly encouraged to get involved with student chapters of professional societies in your discipline. They provide many benefits, such as group meetings that allow you to interact with industry, fellow students, and faculty. If you become a student officer, the leadership experience will be invaluable to your future success. Many student chapters arrange for plant trips so you can learn about the "real world" of engineering. Also, student chapters have social gatherings where you can become better acquainted with your peers.

2.9   THE ENGINEERING DESIGN METHOD

In high school, you probably have been exposed to the **scientific method:**

1. Develop *hypotheses* (possible explanations) of a physical phenomenon.

TABLE 2.2
Internet addresses of major professional societies

AAES	American Association of Engineering Societies	www.aaes.org
NSPE	National Society of Professional Engineers	www.nspe.org
IEEE	The Institute of Electrical and Electronics Engineers	www.ieee.org
ASCE	American Society of Civil Engineers	www.asce.org
ASME	The American Society of Mechanical Engineers	www.asme.org
AIChE	American Institute of Chemical Engineers	www.aiche.org
IIE	Institute of Industrial Engineers	www.iienet.org
AIAA	American Institute of Aeronautics and Astronautics	www.aiaa.org
ACM	Association for Computing Machinery	www.acm.org
AIME	American Institute of Mining, Metallurgical and Petroleum Engineering	www.aimeny.org
ASAE	American Society of Agricultural and Biological Engineers	www.asabe.org
ANS	American Nuclear Society	www.ans.org
BMES	Biomedical Engineering Society	www.bmes.org
MAES	Society of Mexican American Engineers and Scientists	www.maes-natl.org
NSBE	National Society of Black Engineers	www.nsbe.org
SWE	Society of Women Engineers	www.swe.org

2. Design an experiment to critically test the hypotheses.
3. Perform the experiment and analyze the results to determine which hypothesis, if any, is consistent with the experimental data.
4. Generalize the experimental results into a law or theory.
5. Publish the results.

Although engineers use knowledge generated by the scientific method, they do not routinely use the method; that is the domain of scientists. The goals of scientists and engineers are different. Scientists are concerned with discovering what *is,* whereas engineers are concerned with designing what *will be.* To achieve our goals, engineers use the **engineering design method,** which is, briefly stated:

1. Identify and define the problem.
2. Assemble a design team.
3. Identify constraints and criteria for success.
4. Search for solutions.
5. Analyze each potential solution.
6. Choose the "best" solution.
7. Document the solution.
8. Communicate the solution to management.
9. Construct the solution.
10. Verify and evaluate the performance of the solution.

This method is described in much greater detail in Chapter 5, "Introduction to Design."

Your engineering education will focus primarily on **analysis.** The hundreds (or thousands) of homework and exam problems you will work during your studies are all designed to sharpen your analytical skills.

China through the Ages: Walls, Words, and Wells

No discussion of ancient engineering feats is complete without mention of the Great Wall of China. Construction began in the third century B.C., under the rule of the brutal emperor Qin Shi Huang Di (a title meaning "First Divine Autocrat of the Qin Dynasty"). The emperor's goal was to secure China from the murderous Huns of northern Asia. To this end, he forced hundreds of thousands of Chinese peasants, men and women, to leave their homes and fields and join the building effort. Though not completed during Qin's lifetime, the Wall eventually grew to a length of over 2200 miles, including the spurs and branches. If placed in America, it would stretch from New York City to Des Moines, Iowa. Materials and dimensions vary over the entire length, but the Wall is largely constructed of clay bricks, 25 feet thick at the base and rising 30 feet high. Watchtowers are spaced every few hundred yards. The Wall has been rebuilt by various rulers throughout history, even into the 19th century.

Photo © Photodisc/Vol. 89.

Great Wall of China.

Later Chinese engineering accomplishments, though not quite as large, are equally remarkable. In the 1st century A.D., the courtly eunuch Cai Lun concocted paper from tree bark, hemp, rags, and fishnets. Later, development of the printing press in the 9th through the 12th centuries made the Chinese the first publishers as well as the first to circulate printed currency.

Pioneering the chemical industry, engineers of the landlocked Sichuan province in central China collected brine from deep wells for salt production as early as the 11th century A.D. Salt accounted for the bulk of the local economy for over 800 years. Using bamboo cables to drill and bamboo pipes to collect, the wells grew from 100 to 1000 meters deep as technology improved. As early as the 16th century, the Sichuanese learned to store the natural gas that also came from the wells, and used it to fire the brine boilers.

Courtesy of Seth Adelson.

In their analysis of physical systems, engineers use **models.** A model represents the real system of interest. Depending upon the quality of the model, it may, or may not, be an accurate representation of reality.

2.9.1 Qualitative Models

A **qualitative model** is a simple relationship that is easily understood. For example, if you were designing a grandfather clock, the *period* of the pendulum—the time it takes to swing back and forth—would be a critical design issue because the pendulum regulates the clock (Figure 2.3). By observing a swinging rock tied to a string, you may notice that longer strings lengthen the period. A simple relationship such as this is very useful to the engineer; however, it is generally insufficient for rigorous analysis. We usually require more quantitative information. To build the clock, we need to know the exact period for a given pendulum length.

2.9.2 Mathematical Models

Because engineering usually needs quantitative values, we transform these qualitative ideas about string length into mathematical formulas. For small displacement angles θ (less than about 15°), physics tells us that the period P of the pendulum (the time it takes to return to its original starting position) can be calculated by the simple formula

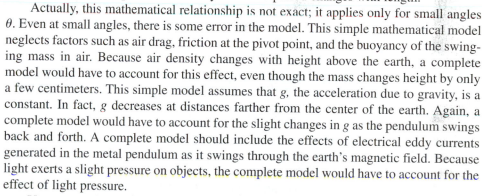

$$P = 2\pi\sqrt{\frac{L}{g}} = \frac{2\pi}{\sqrt{g}}\sqrt{L} = k\sqrt{L} \tag{2-1}$$

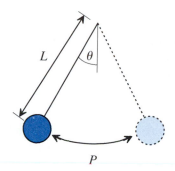

L

θ

P

FIGURE 2.3
Pendulum.

where L is the pendulum length (measured from the pivot point to the center of the pendulum mass), g is the acceleration due to gravity (9.8 m/s^2), and k is a proportionality constant. This relationship tells us "exactly" how the period changes with length.

Actually, this mathematical relationship is not exact; it applies only for small angles θ. Even at small angles, there is some error in the model. This simple mathematical model neglects factors such as air drag, friction at the pivot point, and the buoyancy of the swinging mass in air. Because air density changes with height above the earth, a complete model would have to account for this effect, even though the mass changes height by only a few centimeters. This simple model assumes that g, the acceleration due to gravity, is a constant. In fact, g decreases at distances farther from the center of the earth. Again, a complete model would have to account for the slight changes in g as the pendulum swings back and forth. A complete model should include the effects of electrical eddy currents generated in the metal pendulum as it swings through the earth's magnetic field. Because light exerts a slight pressure on objects, the complete model would have to account for the effect of light pressure.

You can see from this discussion that a complete model of the pendulum is hopelessly complex. Engineers rarely are able to develop complete mathematical models. However, even incomplete mathematical models may be extremely useful for design purposes, so we use them. A good engineer designs the final product so adjustments can be made to correct for minor effects not considered in the model, or to accommodate slight variations in the manufacturing process. In the case of the grandfather clock, the pendulum could have an adjustment screw that slightly changes its length.

Once a mathematical model of the system has been developed, the complete power of mathematics is at the disposal of the engineer to manipulate the mathematical description of the system. Insofar as the mathematical model is a reasonably accurate description of reality, the mathematical manipulations will also result in equations that approximate reality.

2.9.3 Digital Computer Models

Mathematical models may be programmed and solved using digital computers. In our pendulum example, we could write a computer program that calculates the position of the pendulum as time progresses. At each position, we could calculate the air density, the buoyancy forces, the acceleration due to gravity, the light-pressure forces, the air drag, and the pivot-point friction. The computer model would use all of this information to calculate the next position. All of this information would then be recalculated, allowing the next position to be determined. This may sound like a lot of work. It is. The amount of modeling effort expended depends upon how accurately the period must be known. Perhaps it would be better to use the simpler model and, using an adjustment screw on the pendulum, calibrate it against an electronic clock.

2.9.4 Analog Computer Models

Electronic circuits can be configured to simulate physical systems. Before digital computers became widely available, analog computers were frequently used. Today, they are rarely used because digital computers are more versatile and powerful.

2.9.5 Physical Models

Some systems are extremely complex and require physical models. For example, wind tunnel models of the space shuttle were constructed to determine its flight characteristics. Engineers use a physical model of the Mississippi River to understand the effect of silt deposits and rainfall on its flow rate. Chemical engineers build a pilot plant to test a chemical process before the industrial-scale plant is constructed.

2.10 TRAITS OF A SUCCESSFUL ENGINEER

All of us would like to be successful in our engineering careers, because it brings personal fulfillment and financial reward. (For most engineers, financial reward is not the highest priority. Surveys of practicing engineers show that they value exciting and challenging work performed in a pleasant work environment over monetary compensation.) As a student, you may feel that performing well in your engineering courses will guarantee success in the real engineering world. Unfortunately, there are no guarantees in life. Ultimate success is achieved by mastering many traits, of which academic prowess is but one. By mastering the following traits, you will increase your chances of achieving a successful engineering career:

- *Interpersonal skills.* Engineers are typically employed in industry where success is necessarily a group effort. Successful engineers have good interpersonal skills. Not only must they effectively communicate with other highly educated engineers, but also with artisans, who may have substantially less education, or other professionals who are highly educated in other fields (marketing, finance, psychology, etc.).
- *Communication skills.* Although the engineering curriculum emphasizes science and mathematics, some practicing engineers report that they spend up to 80% of their time in oral and written communications. Engineers generate engineering drawings or sketches to describe a new product, be it a machine part, an electronic circuit, or a crude flowchart of new computer code. They document test results in reports. They write memos, manuals, proposals to bid on jobs, and technical papers for trade journals. They give sales presentations to potential clients and make oral presentations at technical meetings. They communicate with the workers who actually build the devices designed

by engineers. They speak at civic groups to educate the public about the impact of their plant on the local economy, or address safety concerns raised by the public.

- *Leadership.* Leadership is one of the most desired skills for success. Good engineering leaders do not follow the herd; rather, they assess the situation and develop a plan to meet the group's objectives. Part of developing good leadership skills is learning how to be a good follower as well.
- *Competence.* Engineers are hired for their knowledge. If their knowledge is faulty, they are of little value to their employer. Performing well in your engineering courses will improve your competence.
- *Logical thinking.* Successful engineers base decisions on reason rather than emotions. Mathematics and science, which are based upon logic and experimentation, provide the foundations of our profession.
- *Quantitative thinking.* Engineering education emphasizes quantitative skills. We transform qualitative ideas into quantitative mathematical models that we use to make informed decisions.
- *Follow-through.* Many engineering projects take years or decades to complete. Engineers have to stay motivated and carry a project through to completion. People who need immediate gratification may be frustrated in many engineering projects.
- *Continuing education.* An undergraduate engineering education is just the beginning of a lifetime of learning. It is impossible for your professors to teach all relevant current knowledge in a 4-year curriculum. Also, over your 40-plus-year career, knowledge will expand dramatically. Unless you stay current, you will quickly become obsolete.
- *Maintaining a professional library.* Throughout your formal education, you will be required to purchase textbooks. Many students sell them after the course is completed. If that book contains useful information related to your career, it is foolish to sell it. Your textbooks should become personalized references with appropriate underlining and notes in the margins that allow you to quickly regain the knowledge years later when you need it. Once you graduate, you should continue purchasing handbooks and specialized books related to your field. Recall that you will be employed for your knowledge, and books are the most ready source of that knowledge.
- *Dependability.* Many industries operate with deadlines. As a student, you also have many deadlines for homework, reports, tests, and so forth. If you hand homework and reports in late, you are developing bad habits that will not serve you well in industry.
- *Honesty.* As much as technical skills are valued in industry, honesty is valued more. An employee who cannot be trusted is of no use to a company.
- *Organization.* Many engineering projects are extremely complex. Think of all the details that had to be coordinated to construct your engineering building. It is composed of thousands of components (beams, ducting, electrical wiring, windows, lights, computer networks, doors, etc.). Because they interact, all those components had to be designed in a coordinated fashion. They had to be ordered from vendors and delivered to the construction site sequentially when they were required. The activities of the contractors had to be coordinated to install each item when it arrived. The engineers had to be organized to construct the building on time and within budget.
- *Common sense.* There are many commonsense aspects of engineering that cannot be taught in the classroom. A lack of common sense can be disastrous. For example, a library was recently built that required pilings to support it on soft ground. (A *piling* is

a vertical rod, generally made from concrete, that goes deep into the ground to support the building that rests on it.) The engineers very carefully and meticulously designed the pilings to support the weight of the building, as they had done many times before. Although the pilings were sufficient to hold the building, the engineers neglected the weight of the books in the library. The pilings were insufficient to carry this additional load, so the library is now slowly sinking into the ground.

- *Curiosity.* Engineers must constantly learn and attempt to understand the world. A successful engineer is always asking, Why?
- *Involvement in the community.* Engineers benefit themselves and their community by being involved with clubs and organizations (Kiwanis, Rotary, etc.). These organizations provide useful community services and also serve as networks for business contacts.
- *Creativity.* From their undergraduate studies, it is easy for engineering students to get a false impression that engineering is not creative. Most courses emphasize **analysis,** in which a problem has already been defined and the "correct" answer is being sought. Although analysis is extremely important in engineering, most engineers also employ **synthesis,** the act of creatively combining smaller parts to form a whole. Synthesis is essential to design, which usually starts with a loosely defined problem for which there are many possible solutions. The creative engineering challenge is to find the best solution to satisfy the project goals (low cost, reliability, functionality, etc.). Many of the technical challenges facing society can be met only with creativity, for if the solutions were obvious, the problems would already be solved.

2.11 CREATIVITY

Imagination is more important than knowledge.
Albert Einstein

If the above quotation is correct, you should expect your engineering education to start with Creativity 101. Although many professors do feel that creativity is important in engineering education, creativity *per se* is not taught. Why is this?

- Some professors feel that creativity is a talent students are born with and cannot be taught. Although each of us has different creative abilities—just as we have different abilities to run the 50-yard dash—each of us *is* creative. Often, all the student needs is to be in an environment in which creativity is expected and fostered.
- Other professors feel that because creativity is hard to grade, it should not be taught. Although it is important to evaluate students, not everything a student does must be subjected to grading. The students' education should be placed above the students' evaluation.
- Other professors would argue that we do not completely understand the creative process, so how could we teach it? Although it is true we do not completely understand creativity, we know enough to foster its development.

Rarely is creativity directly addressed in the engineering classroom. Instead, the primary activity of engineering education is the transfer of knowledge to future generations that was painstakingly gained by past generations. (Given the vast amount of knowledge, this is a Herculean task.) Further, engineering education emphasizes the proper manipulation of knowledge to correctly solve problems. Both these activities support analysis, not synthesis.

TABLE 2.3
Creative professions

Profession	Goals	Constraints
Author	Communication, exploration of emotions, development of characters	Language
Artist	Communication, creation of beauty, experimentation with different media	Visual form
Composer	Communication, creation of new sounds, exploration of potential of each instrument	Musical form
Engineer	Simplicity, increased reliability, improved efficiency, reduced cost, better performance, smaller size, lighter weight, etc.	Physical laws and economics

The "analysis muscles" of an engineering student tend to be well developed and toned. In contrast, their "synthesis muscles" tend to be flabby due to lack of use. Both analysis and synthesis are part of the creative process; engineers cannot be productively creative without possessing and manipulating knowledge. But it is important to realize that if you wish to tone your "synthesis muscles," it may require activities outside the engineering classroom.

Table 2.3 lists some creative professions, of which engineering is one. Although the goals of authors, artists, and composers are many, most have the desire to communicate. However, the constraints placed upon their communication are not severe. The author e.e. cummings is well known for not following grammatical conventions. We have all been to art galleries in which a blob passes for art. The musician John Cage composed a musical piece entitled 4′ 33″ in which the audience listens to random ambient noise (e.g., the air handling system, coughs, etc.) for 4 minutes and 33 seconds.

The goals of engineers differ from those of the other creative professions (Table 2.3). To achieve these goals, we are constrained by physical laws and economics. Unlike other creative professions, we are not free to ignore our constraints. What success would an aerospace engineer achieve by ignoring gravity? Because we must work within constraints to achieve our goals, engineers must exhibit tremendous creativity.

Of those engineering goals listed in Table 2.3, one of the most important is simplicity. Generally, a simple design tends to satisfy the other goals as well. The engineer's desire to achieve simplicity is known as the *KISS* principle: "Keep It Simple, Stupid."

Although the creative process is not completely understood, we present here our own ideas about the origins of creativity. People can crudely be classified into *organized thinkers, disorganized thinkers,* and *creative thinkers.* Imagine we tell each of these individuals that "paper manufacture involves removing lignin (the natural binding agent) from wood, to release cellulose fibers that are then formed into paper sheets." Figures 2.4 through 2.6 show how each thinker might store the information.

The organized thinker has a well-compartmentalized mind. Facts are stored in unique places, so they are easily retrieved when needed. The papermaking fact is stored under "organic chemistry," because lignin and cellulose are organic chemicals.

The disorganized thinker has no structure. Although the information may be stored in multiple places, his mind is so disorganized that the information is hard to retrieve when

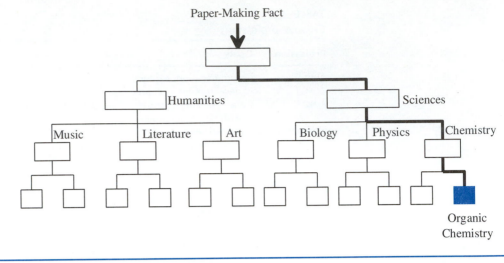

FIGURE 2.4
Papermaking fact stored
by an organized thinker.

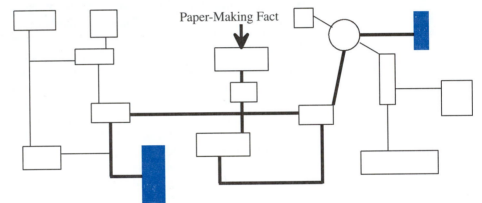

FIGURE 2.5
Papermaking fact stored by
a disorganized thinker.

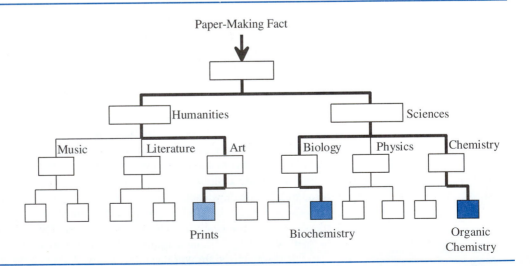

FIGURE 2.6
Papermaking fact stored
by a creative thinker.

needed. The disorganized thinker who needed to recall information about papermaking would not have a clue where to find it.

The creative thinker is a combination of organized and disorganized thinkers. The creative mind is ordered and structured, but information is stored in multiple places so that when the information is needed, there is a higher probability of finding it. When creative people learn, they attempt to make many connections, so the information is stored in different places and is linked in a variety of ways. In the papermaking example, they might store the information under "organic chemistry" because they are organized, but also under "biochemistry" (because lignin and cellulose are made by living organisms) and under "art prints" (because high-quality prints must be printed on "acid-free" paper, which uses special chemistry to remove the lignin).

When an engineer tries to solve a problem, she works at both the conscious and subconscious level (Figure 2.7). The subconscious seeks information that solves a qualitative model of the problem. As long as it finds no solution, the subconscious mind keeps searching the information data banks. Here, we see the advantage of the creative thinker. With information stored in multiple places and connected in useful ways, there is a greater probability that a solution to the qualitative model will be found. When the subconscious finds a solution, it emerges into consciousness. You have certainly experienced this. Perhaps you went to bed with a problem on your mind, and when you woke up, the solution seemingly "popped" into your head. In actuality, the subconscious worked on the problem while you were sleeping, and the solution emerged into your consciousness when you awoke. For engineers, generally what emerges from the subconscious is a potential solution. The actual solution won't be known until the potential solution is analyzed using a quantitative model. If analysis proves the solution, then the engineer has cause for celebration; she has solved the problem.

Most of your engineering education will focus on analysis, the final step in the problem-solving process. However, unless your subconscious is trained, you won't have good potential solutions to analyze. Notice that the subconscious requires a qualitative model. A good engineer develops a "feeling" for numbers and processes and

The cartoons of Rube Goldberg are well-known for accomplishing a simple task in an overly complex manner, thus violating the KISS principle.

INVENTIONS OF PROFESSOR LUCIFER BUTTS

THE PROFESSOR'S BRAIN TOSSES OFF HIS LATEST ANTI-FLOOR WALKING PARAPHERNALIA — Pull string (A) which discharges pistol (B) and bullet (C) hits switch on electric stove (D), warming pot of milk (E). Vapor from milk melts candle (F) which drips on handle of pot causing it to upset and spill milk down trough (G) and into can (H). Weight bears down on lever (I) pulling string (J) which brings nursing nipple (K) within baby's reach.

In the meantime baby's yelling has awakened two pet crows (L & M) and they discover rubber worm (N) which they proceed to eat. Unable to masticate it, they pull it back and forth causing cradle to rock and put baby to sleep.

Put cotton in your ears so you will not be bothered if baby wakes again —

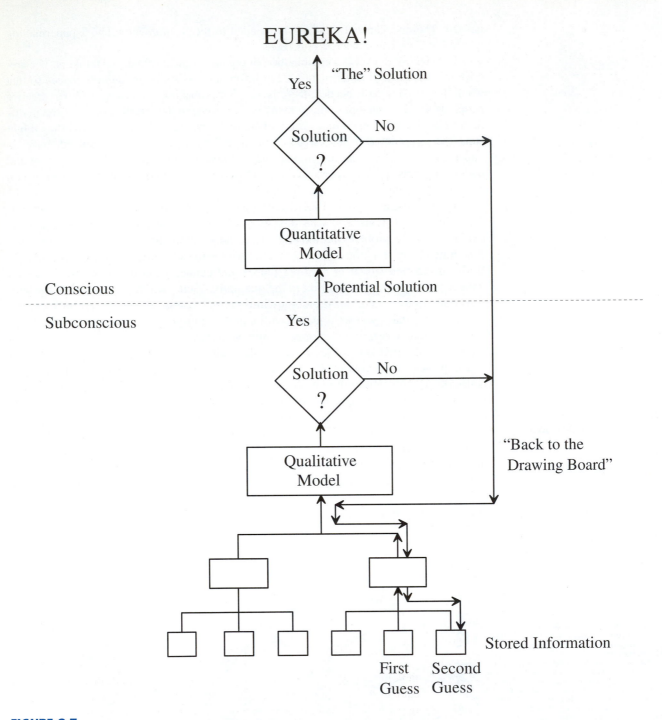

FIGURE 2.7
The problem-solving process.

often does not have to feed mathematical formulas to get answers. Developing a feeling for numbers will also help your analysis skills, as it provides an essential check on your calculated answers.

2.12 TRAITS OF A CREATIVE ENGINEER

The following list describes some traits of a creative engineer:

- *Stick-to-it-iveness.* Producing creative solutions to problems requires unbridled commitment. There are always problems along the way. A successful creative engineer does not give up. Thomas Alva Edison said that "Genius is 1 percent inspiration and 99 percent perspiration."
- *Asks why.* A creative engineer is curious about the world and is constantly seeking understanding. By asking why, the creative engineer can learn how other creative engineers solved problems.
- *Is never satisfied.* A creative engineer goes through life asking, How could I do this better? Rather than complaining about a stoplight that stops his car at midnight when there is no other traffic, the creative engineer would say, How could I develop a sensor that detects my car and turns the light green?
- *Learns from accidents.* Many great technical discoveries were made by accident (e.g., Teflon). Instead of being single-minded and narrow, be sensitive to the unexpected.
- *Makes analogies.* Recall that problem solving is an iterative process that largely involves chance (Figure 2.7). By having rich interconnections, a creative engineer increases the chance of finding a solution. We obtain rich interconnections by making analogies during learning so information is stored in multiple places.
- *Generalizes.* When a specific fact is learned, a creative engineer seeks to generalize that information to generate rich interconnections.
- *Develops qualitative and quantitative understanding.* As you study engineering, develop not only quantitative analytical skills, but also qualitative understanding. Get a feeling for the numbers and processes, because that is what your subconscious needs for its qualitative model.
- *Has good visualization skills.* Many creative solutions involve three-dimensional visualization. Often, the solution can be obtained by rearranging components, turning them around, or duplicating them.
- *Has good drawing skills.* Drawings or sketches are the fastest way by far to communicate spatial relationships, sizes, order of operations, and many other ideas. By accurately communicating through engineering graphics and sketches, an engineer can pass her ideas easily and concisely to her colleagues, or with a little explanation, to non-engineers.
- *Possesses unbounded thinking.* Very few of us are trained in general engineering. Most of us are trained in an engineering discipline. If we restrict our thinking to a narrowly defined discipline, we will miss many potential solutions. Perhaps *the* solution requires the combined knowledge of mechanical, electrical, and chemical engineering. Although it is unreasonable that we be expert in all engineering disciplines, each of us should develop enough knowledge to hold intelligent conversations with those in other disciplines.

- *Has broad interests.* A creative engineer must be happy. This requires balancing intellectual, emotional, and physical needs. Engineering education emphasizes your intellectual development; you are responsible for developing your emotional and physical skills by socializing with friends, having a stimulating hobby (e.g., music, art, literature), and working out.
- *Collects obscure information.* Easy problems can be solved with commonly available information. The hard problems often require obscure information.
- *Works with nature, not against it.* Do not enter a problem with preconceived notions about how it must be solved. Nature will often guide you through the solution if you are attentive to its whispers.
- *Keeps an engineering "toolbox."* An engineering "toolbox" is filled with simple qualitative relationships needed by the qualitative model in the subconscious. These simple qualitative relationships may be the distilled wisdom from quantitative engineering analysis. The following sections describe a few "tools." As you progress through your career, you will need a large toolbox to hold all the tools you acquire from your experience.

2.12.1 Cube-Square Law

An example of information an engineer may store in her toolbox is the **cube-square law.** The cube-square law says that as an object gets smaller, its volume decreases much faster than its area. Therefore, the surface-area-to-volume ratio increases with smaller objects.

To illustrate this law, imagine our object is a sphere (Figure 2.8). The surface area A is

$$A = 4\pi r^2 \tag{2-2}$$

and the volume V is

$$V = \frac{4}{3}\pi r^3 \tag{2-3}$$

The surface-area-to-volume ratio is

$$\frac{A}{V} = \frac{4\pi r^2}{\frac{4}{3}\pi r^3} = \frac{3}{r} \tag{2-4}$$

This equation says that the area-to-volume ratio increases as the radius decreases.

The cube-square law is one of the most overarching laws in nature, as shown by the following examples:

Example 2.1. Imagine you work in a cannonball factory that casts cannonballs from molten metal and cools them in air. From the cube-square law, you know that smaller cannonballs will cool much faster than large cannonballs, because the rate of heat loss is affected by the surface area but the total amount of heat loss is determined by the cannonball volume.

Example 2.2. Imagine that you must select the most energy-efficient method to fly 500 passengers from New York to Paris. You could charter five 100-passenger planes or one 500-passenger plane. Fuel is primarily required to overcome air drag, which is

FIGURE 2.8

Illustration of the cube-square law using a sphere.

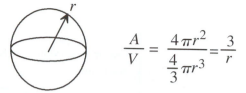

$$\frac{A}{V} = \frac{4\pi r^2}{\frac{4}{3}\pi r^3} = \frac{3}{r}$$

dictated by the plane surface area. Passenger capacity is determined by the plane volume. To improve fuel economy, you want a small surface area relative to the volume, so one large plane is better than five smaller planes.

Example 2.3. You want to store 50,000 gallons of diesel fuel. You are contemplating purchasing one 50,000 gallon tank or five 10,000 gallon tanks. Tank vendors charge for the metal, not the air in the tank; therefore, tank cost is mostly determined by the surface area. Because a large tank has less surface area per volume, it will be less expensive to purchase one large tank rather than five smaller tanks.

Example 2.4. A whale has limited surface area relative to its volume. Metabolic energy generated within the whale's volume must be eliminated to the surrounding fluid via the whale's surface area. When swimming in water, there is no problem transferring the heat, because water has good heat transfer properties. But when a whale becomes beached on land, it is surrounded by air, which has poor heat transfer properties. The whale literally cooks, because it cannot transfer metabolic energy to the surroundings.

Example 2.5. Normally, coal burns safely at a controlled rate within a furnace. However, in mining operations, fine coal dust is produced, which can burn explosively because there is so much surface area relative to its volume. This presents a significant safety hazard against which numerous precautions are taken.

Example 2.6. The mass of an animal is dictated by its volume, but leg strength is determined by its cross-sectional area. If a deer were scaled up to the size of an elephant, its long slender legs would snap. That is why an elephant must have short, thick legs.

2.12.2 Law of Diminishing Returns

Many engineering systems exhibit the behavior shown in Figure 2.9. With small inputs, the output increases linearly, but with larger inputs, the output levels off. There are "diminishing returns" in the nonlinear region, meaning there is not as much output per unit of input.

Example 2.7. The more time an engineer spends designing a product, the better the product becomes. In this example, time is the input and product quality is the output. In the early stages of the design, product quality improves dramatically with each additional hour spent on the project. However, at some point, an additional hour spent does

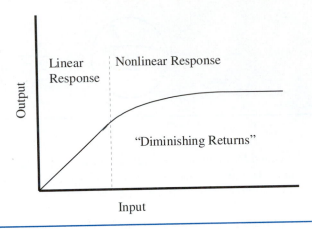

FIGURE 2.9
The **law of diminishing returns.**

not improve the design much more. Once this occurs, the engineer is getting diminishing returns on his effort.

Example 2.8.   Crop production improves as fertilizer is applied to a farmer's field. In this example, fertilizer is the input and crops are the output. Low fertilizer additions improve crop production dramatically *per unit of fertilizer added*. Although high fertilizer additions do increase crop production, the improvement *per unit of fertilizer added* is low. An agricultural engineer designing a crop production system would realize diminishing returns at high fertilizer additions.

2.12.3 Put Material Where Stresses Are Greatest

When designing a structure or product, engineers minimize the amount of materials used. This is an important objective for many reasons:

- The cost of most products is directly related to the amount of material in them; therefore, by reducing the amount of material, the product will be less expensive.
- The size of the support columns on the lower levels of a building are proportional to the mass of the materials used in the upper levels. If the upper levels are heavy, the support columns become so large that there is little usable space available.
- Fuel requirements increase as a vehicle becomes more massive. Reducing materials from an automobile improves fuel economy. Similarly, a very powerful engine is required to accelerate a heavy automobile, whereas a less powerful engine can accelerate a light automobile.
- The cost of lifting objects into earth orbit is about $20,000 per kilogram. Lightening the load reduces launch costs.
- Portable products (e.g., notebook computers) must be lightweight to be accepted by the public.

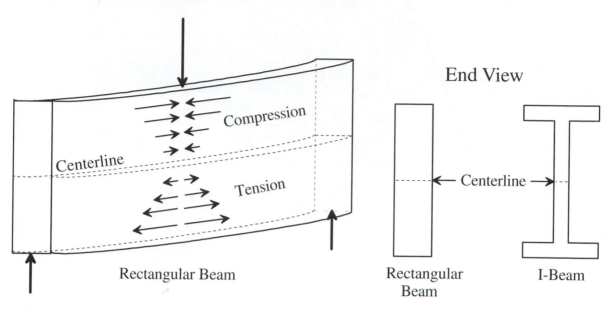

FIGURE 2.10
Loading of a rectangular beam. End view of a rectangular beam and I-beam.

The weight of a product can be reduced by putting the materials where the stresses are, as illustrated by the following examples:

Example 2.9. Figure 2.10 shows the stresses in a rectangular beam that is loaded in the middle and supported on the two ends. The material on the top edge is in compression, the material on the bottom edge is in tension, and the material on the centerline has no stress. By removing material from the center area (where there is little stress) and adding it to the edges, the beam becomes much stronger. This is the design principle for an I-beam, which is commonly used in large buildings.

Example 2.10. Figure 2.11 shows a dam holding back a lake. As the water depth increases, so does the pressure, which puts more stress at the bottom of the dam. To counteract these greater stresses, the dam is thicker at the bottom.

2.13   SUMMARY

Engineers are individuals who combine knowledge of science, mathematics, and economics to solve technical problems that confront society. As civilization has progressed and become more technological, the engineers' impact on society has increased.

    Engineers are part of a technology team that includes scientists, technologists, technicians, and artisans. Historically, various disciplines within engineering have evolved (e.g., civil, mechanical, industrial). Regardless of their discipline, engineers fulfill many functions (research, design, sales, etc.). Because of engineers' importance to society, their education is regulated by ABET and professional licenses are granted

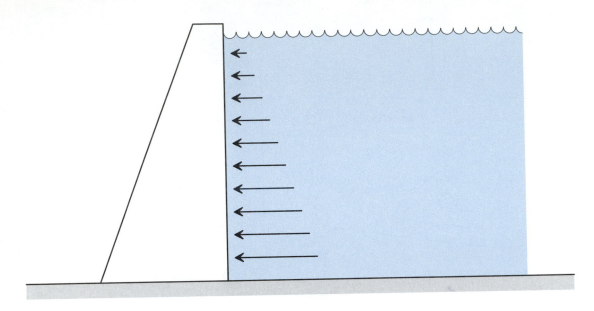

FIGURE 2.11
Stresses on a dam.

by states. There are many engineering professional societies serving a variety of roles, such as providing continuing education courses and publishing technical journals.

In meeting the needs of society, engineers use the engineering design method. An important step in the engineering design method is to formulate models of reality. These models can range from simple qualitative relationships to detailed quantitative codes in digital computers.

To be successful, engineers need to cultivate many traits, such as competence and communication skills. Among the more important skills is creativity, which is needed to solve the more difficult problems faced by society. The creative process involves an interplay between qualitative models that are understood by the subconscious and quantitative models understood by the conscious. These qualitative models may be viewed as tools that engineers keep in their "toolbox" to help guide their creativity in productive directions.

Further Readings

Beakley, G. C., and H. W. Leach. *Engineering: An Introduction to a Creative Profession,* 4th ed. New York: Macmillan, 1983.

Eide, A. R., R. D. Jenison, L. L. Northup, and S. K. Mickelson. *Engineering Fundamentals and Problem Solving,* 5th ed. New York: McGraw-Hill, 2008.

Florman, S. C. *The Existential Pleasures of Engineering,* 2nd ed. New York: St. Martin's Press, 1996.

Hickman, L. A. *Technology as a Human Affair.* New York: McGraw-Hill, 1990.

Petroski, H. *Beyond Engineering.* New York: St. Martin's Press, 1986.

———. *To Engineer Is Human: The Role of Failure in Successful Design.* New York: Vintage Books, 1992.

Smith, R. J., B. R. Butler, and W. K. LeBold. *Engineering as a Career,* 4th ed. New York: McGraw-Hill, 1983.

Wright, P. H. *Introduction to Engineering,* 3rd rev. ed. New York: Wiley, 2002.

PROBLEMS

2.1 Interview an engineer and write a one-page report about his or her experiences in college and industry, including joys and frustrations.

2.2 Interview an engineer who performs one of the job functions listed in the chapter and write one paragraph about his or her experiences.

2.3 Identify five companies you might wish to work for, and explain why you would like to work for them.

2.4 Pretend that next week you are going for a job interview with a company of your choice. Write a one-page report on this company, which will help you prepare for your interview.

2.5 Identify a topic or problem that requires the cooperation of two (or more) engineering disciplines.

2.6 Explain why engineers mostly work in teams.

2.7 For each of the following industries, list all the engineering disciplines they employ:
(a) nuclear power plant
(b) automobile manufacturer
(c) airplane manufacturer
(d) building construction
(e) computer manufacturer
(f) computer chip manufacturer
(g) oil refinery

2.8 Describe a world that lacks one of the following engineering disciplines:
(a) mechanical
(b) civil
(c) chemical
(d) computer
(e) electrical

2.9 Look around the room you are in and identify how engineers were involved in the creation of 10 things.

2.10 Write a one-page biography on a famous engineer. You may choose your own, or select from the following list:
(a) James Watt
(b) Johann Gutenberg
(c) Paul MacCready
(d) Herbert Hoover
(e) Thomas Edison
(f) Bill Mullholland
(g) Burt Rutan
(h) Lee Iacocca
(i) John Roebling
(j) Wilbur and Orville Wright
(k) Wilson Greatbatch
(l) Robert Goddard
(m) Henry Ford
(n) Hedy Lamar
(o) Charles Goodyear
(p) Cyrus McCormick

2.11 Write a one-page report on the relationship between engineering and business based upon the following examples:
(a) James Watt/Matthew Boulton
(b) Nikola Tesla/George Westinghouse
(c) Wilbur and Orville Wright/Glenn Curtiss

2.12 Report on the role engineers played in averting the Apollo 13 disaster.

2.13 Write a one-page report describing the Einstein/Szilard refrigerator and how it was created.

Glossary

analog An electronic device that uses continuous values of voltages and currents.

analog computer model An analog electronic circuit that simulates a physical system.

analysis The process of defining and seeking an answer to a problem.

cube-square law A law of nature that says as an object gets smaller, its volume decreases much faster than its area; therefore, the surface-area-to-volume ratio increases with smaller objects.

digital An electronic device that uses only discrete voltages and currents.

digital computer model A program in a digital computer that simulates a physical system.

engineer An individual who combines knowledge of science, mathematics, and economics to solve technical problems that confront society.

engineering design method A procedure of synthesis, analysis, communication, and implementation used to design solutions to problems.

law of diminishing returns The concept that with small inputs, the output increases linearly, but with larger inputs, the output levels off.

mathematical model Mathematical formulas that describe a physical system.

model A representation of a real physical system.

physical model A physical representation of a complex system.

qualitative model A non-numerical description of a physical system.

synthesis The act of creatively combining smaller parts to form a coherent whole.

CHAPTER 3

Engineering Ethics

Engineering is a *profession,* similar to law, medicine, dentistry, and pharmacy. A distinguishing feature of all these professions is that their practitioners are highly educated. Engineers are hired by clients (and employers) specifically for their specialized expertise. Generally, the client knows less about the subject than the engineer. Therefore, engineers have ethical obligations to the clients, because the client often cannot assess the quality of the engineer's technical advice. These obligations are part of **engineering ethics,** the set of behavioral standards that all engineers are expected to follow. Engineering ethics are an extension of the ethical standards we all have as human beings.

Engineers have a long tradition of ethical behavior that is widely recognized. Public opinion polls consistently list engineering among the most ethical professions.

3.1 INTERACTION RULES

Engineers rarely work as lone individuals; we generally work in teams. Further, the products of our labor—automobiles, roads, chemical plants, computers—impact society as a whole. Therefore, we need a set of **interaction rules** outlining the expected sets of behavior between the engineer, other individuals, and society as a whole (Figure 3.1). The interaction rules go both ways: the engineer has obligations to society (e.g., to be honest, unbiased, hardworking, careful) and society has obligations to the engineer (e.g., to pay for work performed, to protect intellectual property).

Interaction rules can be classified as etiquette, law, morals, and ethics.

3.1.1 Etiquette

Etiquette consists of codes of behavior and courtesy. It addresses such issues as how many forks to place on the dinner table, proper dress at weddings, seating arrangements, and invitations to parties. Although we generally learn these rules from our everyday experience, they have been codified in various books.

The rules of etiquette are often arbitrary, and they evolve rapidly. For example, in the past, it was common for women to wear white gloves at formal functions; now, this is rarely done. The consequences of violating rules of etiquette are generally not severe. Although a faux pas may cause those "in the know" to snicker, it does not result in jail time. In some cases, etiquette can have important impacts. The peace talks during the

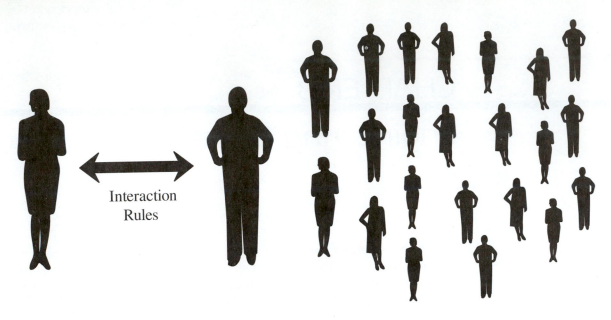

Interaction
Rules

FIGURE 3.1
Interaction rules between other individuals and society.

Vietnam War were mired down in a discussion about the proper shape of the negotiating table; many lives were lost during the delay.

Within the engineering world, proper etiquette is manifested by showing proper respect to employers and clients, not embarrassing colleagues, answering the phone in a professional manner, and so forth.

3.1.2 Law

Law is a system of rules established by authority, society, or custom. Unlike etiquette, violations of law carry penalties such as imprisonment, fines, community service, death, dismemberment, or banishment. Each society has its own consequences for law violations. In Middle Eastern societies, robbers may have a hand amputated, whereas Western society favors imprisonment.

Because severe penalties may result from violating the law, the law must be clearly identifiable so everyone can know the boundaries separating legal and illegal behavior. In some cases, this requirement may lead to seemingly arbitrary laws. For example, it is illegal for a 15-year-old to drive in most states, whereas 16-year-olds usually can. This is somewhat arbitrary, as a mature 14-year-old may be capable of driving but an immature 18-year-old may not. Because we have no test to quantify the maturity level needed for driving, society relies on age as a nonsubjective (although imperfect) measure of maturity. Similarly, the United States specifies 21 as the responsible drinking age. Legal driving ages and drinking ages vary between countries, another indication that an exact age is somewhat arbitrary.

Legal rights are "just claims" given to all humans within a government's jurisdiction. Most governments grant their citizens rights through a constitution. For example, Amendment VIII of the U.S. Constitution protects citizens from "cruel and unusual punishments." A citizen need not perform acts of kindness to earn this right, nor can it be removed if he or she engages in crime.

3.1.3 Morals

Morals are accepted standards of right and wrong that are usually applied to personal behavior. We derive moral standards from our parents, religious background, friends, and the media (television, movies, books, the Internet, and music). Many moral codes are recorded in religious writings. Despite the wide variety of cultures and religions in the world, there is agreement on many moral standards. Most cultures consider murder and stealing to be immoral behavior. From the view of cultural evolution, we can say there is a strong selective pressure against these behaviors. Societies that did not develop moral codes against these behaviors degenerated into anarchy and disappeared.

For some behaviors, there is not universal agreement as to whether they are immoral. Activities such as gambling, dancing, and consumption of alcohol, meat, coffee, and cigarettes are considered immoral in some cultures and religions, but not in others. From the view of cultural evolution, we can say there is not a strong selective pressure against these activities. Societies can maintain viability while tolerating these activities, although some people would argue that a society that bans them is stronger.

Moral rights are "just claims" that belong to all humans, regardless of whether these rights are recognized by government. Civilization recognizes that simply being a human endows us with rights; we need not do anything to earn these rights. For example, most of the civilized world believes that because prisoners are human beings, they should not be tortured, regardless of the cruelty of their crime. This moral right has been codified as a legal right in the U.S. Constitution (Amendment VIII).

Moral rights can be a source of controversy. For example, do we all have a right to health care? If so, how much? Some people believe that moral rights should be extended to nonhuman species. If this concept is widely adopted, it will limit some engineering activities. For example, the construction of dams and drainage of swamps can dramatically impact the survival of animal and plant species.

3.1.4 Ethics

Ethics consists of general and abstract concepts of right and wrong behavior culled from philosophy, theology, and professional societies. Because professions draw their members from many cultures and religions, their ethical standards must be secular. Most professional societies have a formal code of ethics to guide their members. Appendix B shows the National Society of Professional Engineers Code of Ethics.

3.1.5 Comparison of Interaction Rules

From the above discussion, you can see we have a complex web of interaction rules governing our behavior. In some cases, all the interaction rules agree. For example, murder is illegal, immoral, unethical, a violation of human rights, and certainly bad etiquette.

In general, lawmakers try to formulate laws that are consistent with morality. However, there can be conflicts between the law and morality for the following reasons:

• The legal system has not considered the situation.

> **Example 3.1.** A chemical company develops a new process that has a waste by-product. Its own internal studies show this by-product to be extremely carcinogenic; however, it is not on the government list of banned chemicals because it is so new. If this company releases large amounts of this carcinogen into the environment, it would not violate the law. However, we would all agree this is immoral behavior.

• Encoding some moral standards into law would be unenforceable.

> **Example 3.2.** Some moral codes ban alcohol. During Prohibition, this ban was enacted into U.S. law. However, it was not enforceable and it created more problems than it solved.

• Laws must be impartial and treat everyone the same.

> **Example 3.3.** Government self-regulations require that all purchases be made through purchasing agents. A government engineer wishes to purchase a used alternator at the local junkyard to perform a quick experiment. Obtaining this part through the purchasing agent is unworkable—it is not readily identifiable by a catalog number, and it takes too long to process the order. The engineer decides to purchase the used alternator with his own funds and reimburse himself with office supplies of equivalent value.
>
> This is not immoral behavior because no theft was involved; however, it is illegal. The government cannot make a regulation stating that "honest employees are entitled to reimburse themselves with office supplies; dishonest employees are required to use the purchasing agent." The law must be impartial and treat every government employee as though he or she were dishonest.

• Laws must govern observable behavior.

> **Example 3.4.**   In some moral codes, to think a bad thought is equivalent to having performed it. How could a law be written that bans bad thoughts?

• Laws may be enacted by immoral regimes.

> **Example 3.5.** Nazi law forbade hiding Jews during World War II, but many considered it their moral obligation to break the law and hide innocent Jews.

Morality may also conflict with legal rights. U.S. citizens have the legal right to free speech. Therefore, we have the legal right to tell racist jokes; however, many people consider this to be immoral. Similarly, although the legal rights of pornography publishers have been protected by freedom of the press, many consider this to be immoral activity.

3.2 SETTLING CONFLICTS

A major purpose of interaction rules is to avoid conflicts between members of society. For example, a law tells us on which side of the road to drive. Without it, there would be many lethal conflicts.

Inevitably, human interactions result in conflicts. To settle a conflict, it is necessary to discern its source, which may result from moral issues, conceptual issues, applications issues, and factual issues.

3.2.1 Moral Issues

A **moral issue** is involved if the issue can be resolved only by making a moral decision. For example, when automobiles were first introduced onto roads, a moral decision had to be made whether limits should be placed upon their speeds. One side of the issue would argue that motorists should go as fast as they wish either for their own pleasure or to save time for their business. The other side of the issue would argue that excessive speeds place other motorists and pedestrians at risk. Clearly, moral considerations favor speed limits, for peoples' lives are more valued than the pleasures or business interests of speed-seeking motorists.

3.2.2 Conceptual Issues

A **conceptual issue** arises when the morality of an action is agreed upon, but there is uncertainty about how it should be codified into a clearly defined law, rule, or policy. For example, society has agreed that innocent motorists should be protected from other motorists who go too fast. The conceptual issue is: What speed is too fast? To resolve the conceptual issue, *too fast* may be defined as "highway speeds that exceed 55 mph under favorable driving conditions or speeds likely to result in an accident under adverse driving conditions such as fog, snow, ice, or rain."

3.2.3 Application Issues

An **application issue** results when it is unclear if a particular act violates a law, rule, or policy. Suppose a motorist is going 50 mph on a road posted for 55 mph. During a light rain, the motorist skids off the road and has an accident. The police officer who arrives at the accident scene must decide whether or not to cite the motorist for excessive speed; in other words, the application issue is whether the light rain qualifies as an adverse driving condition.

3.2.4 Factual Issues

A **factual issue** arises when there is uncertainty about morally relevant facts. It can usually be resolved by acquiring more information. If a motorist is stopped by a police officer for going 60 mph in a 55-mph zone, and the motorist claims she was going 55 mph, the conflict can be resolved by getting more data. For example, the motorist might be able to show that the police radar gun is out of calibration by 5 mph.

Notice that the preceding issues were arranged from the most abstract to the most concrete. Factual issues are more clearly defined and can generally be resolved regardless of upbringing and cultural background. In contrast, moral issues are often hard to define, and their resolution may depend upon upbringing and cultural background. As a result, moral issues can be difficult to resolve. They are addressed by applying moral theories to help the decision-making process.

3.3   MORAL THEORIES

As much as we would like to have a "moral algorithm" that always leads us to the correct answer, such an algorithm does not exist. If it did, we could program computers to make moral and ethical decisions for us. Instead, we have **moral theories** that provide a framework for making moral and ethical decisions. Sometimes these different theories lead to different answers, but often they lead to the same answer.

To illustrate moral theories, consider this example: A civil engineer works for the city as a building inspector. As a large building is being erected, this engineer is responsible for ensuring it is built according to the city code. (The code protects the public by specifying proper construction materials and methods suitable for a particular city. For example, buildings in San Francisco are constructed according to a code that allows them to withstand earthquakes.) The building inspector is offered a $10,000 bribe to overlook some shoddy construction that would cost the contractor $50,000 to correct. Should the engineer accept the bribe?

It does not take a great moral theorist to determine that the answer is no. Each moral theory arrives at this answer through slightly different paths.

3.3.1   Ethical Egoism

Ethical egoism is a moral theory stating that an act is moral provided you act in your enlightened self-interest. For example, if a mugger were to attack you with a knife and you killed that mugger in self-defense, this would not be immoral.

Societies that structure themselves to harness our natural desire to act in self-interest are more successful. (The collapse of communism attests to this.) However, ethical egoism is not a license for selfish behavior. In the long term, selfish behavior is not rewarded; selfish people have few friends and are not likely to be promoted in a company.

In our example, if the building inspector were to take the bribe, there is always a chance he or she would be caught. Imprisonment and loss of job are certainly not worth $10,000. Therefore, it is in his or her self-interest not to take the bribe.

Not all ethical and moral issues involve a single individual. In cases where many people or a society are involved, we must consider the broader moral theories of utilitarianism and rights analysis.

3.3.2   Utilitarianism

According to **utilitarianism,** moral activities are those that create the most good for the most people. This moral theory attempts to optimize the **happiness objective function,** such that

Happiness objective function

$$= \sum_i (\text{benefit})_i (\text{importance})_i - \sum_j (\text{harm})_j (\text{importance})_j$$

Those actions that increase benefits and reduce harm are considered best.

To perform utilitarian analysis:

1. Determine the target audience (e.g., an individual, a company, or a society).
2. For each action, determine the harms, benefits, and importance to the target audience.

3. Evaluate the happiness objective function for each action.
4. Select the action that maximizes the happiness objective function.

In our building-inspector example, if he or she were to apply this moral theory, the $10,000 bribe would be counted as a benefit, the $50,000 savings to the contractor would be counted as a benefit, but the deaths from the building collapse would qualify as a harm. The harm so overwhelms the benefits that the correct action is clear.

Utilitarianism is very logical and appealing to engineers. Nonetheless, it has some problems: (1) it implies we have enough knowledge to evaluate the happiness objective function, (2) value judgments are required to assess the importance of each harm/benefit, and (3) it may lead to injustice for individuals.

EXAMPLE 3.6

Problem Statement: Boonville is a small agricultural community located along a major highway. Because farming is not very profitable, there is a lot of poverty. The schools are poor; many children drop out and never amount to much. There is no hospital in Boonville, so health care is poor.

A major trucking company decides it would like to locate a distribution center in Boonville because it is strategically located on the highway. The town council must decide whether to allow the company to locate there. During the public debate, the yea-sayers emphasize that the company will bring many jobs to the community. The resulting tax base will allow them to improve the school system and build a hospital. The nay-sayers observe that the increased truck traffic will probably cause one additional highway death every 5 years. They also point out that there will be a lot of noise and pollution.

Calculate the happiness objective function to determine an ethical course of action.

Solution: The happiness objective function is

$$\text{Happiness objective function} = (\text{good schools})(\text{importance})$$

$$+ (\text{hospital})(\text{importance})$$

$$+ (\text{employment})(\text{importance})$$

$$- (\text{death})(\text{importance})$$

$$- (\text{noise})(\text{importance})$$

$$- (\text{pollution})(\text{importance})$$

When applied to the community, the equation is

$$\text{Happiness objective function} = (8)(10) + (10)(7) + (5)(20) - (50)(2)$$

$$- (3)(1) - (7)(2)$$

$$= + 133 \Rightarrow \text{Do it!}$$

When applied to the family whose child was killed by the truck, the equation is

$$\text{Happiness objective function} = (8)(10) + (10)(7) + (5)(20)$$

$$- (50)(1{,}000{,}000) - (3)(1) - (7)(2)$$

$$= - 49{,}999{,}767 \Rightarrow \text{Don't do it!}$$

Logically, utilitarian analysis is applied only to the entire community affected by the decision, and not subsets of the community. However, here we applied this analysis to a subset of the community (the family) to show that utilitarianism can benefit society as a whole at the expense of individuals. Rights analysis, explained next, attempts to correct this failing of utilitarianism.

3.3.3 Rights Analysis

According to **rights analysis,** moral actions are those that equally respect each human being. This is often summarized in the *Golden Rule:* Do unto others as you would have them do unto you. Many cultures use the Golden Rule; however, it does not work in every case. If it were strictly followed, a manager (who of course would not want to be laid off) could not lay off workers even if it were required for the health of the company.

As another example of Golden Rule failure, consider an Italian foreman who likes to tell Polish jokes. His Polish subordinates are offended and complain to him. He counters that he doesn't mind Italian jokes, and proceeds to tell one.

To solve this problem, we could formulate the *Revised Golden Rule:* Do unto others as *they* would have done unto *them.* The Revised Golden Rule would ask the foreman to put himself in the shoes of his subordinates. Feeling the pain his subordinates experience from the jokes, he would stop his offensive behavior even though he personally is not offended by the jokes. Even the Revised Golden Rule cannot be applied universally. If it were, a judge would be unable to sentence criminals to jail, because the criminals certainly do not want to go to jail.

Because not all rights are equally important, a hierarchy has been established. They are listed below from most important to least important:

1. Right to life, physical integrity, and mental health.
2. Right to maintain one's level of purposeful fulfillment (e.g., right not to be deceived, cheated, robbed, or defamed).
3. Right to increase one's level of purposeful fulfillment (e.g., right to self-respect, to nondiscrimination, and to acquire property).

To perform rights analysis:

1. Determine the target audience.
2. Evaluate the seriousness of the rights infringement according to the above list.
3. Choose the course of action that imposes the least-serious rights infringement.

In our example of the building inspector who was offered a bribe, the correct action to take could be determined via rights analysis. Accepting the $10,000 bribe may lead to a more fulfilling life, but this is subordinate to the rights of those persons who may be killed if the building collapses.

Example 3.7.   Suppose a kidnapper has taken a person hostage and threatens to kill him. The police become involved and lay a trap that involves deception. Here we have a conflict between the rights of the hostage not to be killed and the rights of the kidnapper not to be lied to. Clearly, the rights of the hostage take precedence over the rights of the kidnapper.

3.3.4 Making Moral Decisions When Moral Theories Diverge

In our building-inspector example, we determined he or she should not take the bribe regardless of the moral theory we applied. This is an example of **convergence.** However, this is not always the case. Sometimes, the moral theories do not agree—they **diverge.**

When applied to society, utilitarianism represents one extreme: Do the most good for society regardless of the consequences to the individual. Rights analysis represents the other extreme, in which individual rights are protected regardless of the impact on society. Society must determine how it will strike a balance between these two extremes.

To illustrate how moral theories can diverge, consider highway construction. Engineers decide on the most efficient route between two points that reduces construction costs and allows motorists to efficiently travel between population centers. Often, the most efficient route goes through some homes. The government will condemn those homes under **"eminent domain"** and reimburse the homeowners according to fair market price. The rights of the individual homeowners are violated. Perhaps they have strong emotional ties to their homes and do not want to sell. However, society benefits by constructing the road, because people can move more quickly between population centers, trucking costs are reduced, and fuel is saved. In this case, society has chosen the utilitarian approach.

As another example of diverging moral theories, consider a situation where a sickly brother has a rare disease that will certainly be fatal if he does not receive a kidney transplant. His healthy brother has a closely matching tissue type, so the transplant would be successful. No other relatives have closely matching tissue types, so their transplanted kidneys would fail. The healthy brother never liked his sickly brother and refuses to give him the kidney. The utilitarian approach would forcibly demand that the healthy brother donate a kidney to his sickly brother, as total happiness is greater with this option. The sickly brother would be helped much more than the healthy brother would be harmed. In contrast, rights analysis would honor the right of the healthy brother not to be dismembered. In this case, society has chosen the rights analysis approach.

Although there are no algorithms to tell us exactly what to do, a reasonable approach to making moral decisions when moral theories diverge is to use utilitarianism unless an individual's rights are seriously violated.

3.4   THE ETHICAL ENGINEER

Most professional societies have prepared ethical codes for their members. The purpose of these codes is to provide guidance to engineers on ethical behavior. A distillation of these codes provides the following guidelines:

1. Protect the public safety, health, and welfare.
2. Perform duties only in areas of competence.
3. Be truthful and objective.
4. Behave in an honorable and dignified manner.
5. Continue learning to sharpen technical skills.
6. Provide honest hard work to employers or clients.
7. Inform the proper authorities of harmful, dangerous, or illegal activities.
8. Be involved with civic and community affairs.
9. Protect the environment. [Only in a few codes.]
10. Do not accept bribes or gifts that would interfere with engineering judgment.
11. Protect confidential information of employer or client.
12. Avoid conflicts of interest.

A **conflict of interest** is a situation in which an engineer's loyalties and obligations may be compromised because of self-interest or other loyalties and obligations. This can lead to biased judgments. Suppose an engineer were responsible for selecting bearings for an engine her employer is constructing. It so happens that her father owns a bearing company that the engineer will inherit when her father dies. This situation makes it very difficult for the engineer to make an unbiased selection of bearings because she has an obvious conflict of interest. She and her family would benefit by selecting her father's bearings. Even if her father's bearings are the best ones for the job, the selection of these bearings gives the *appearance* of impropriety. Therefore, the situation must be avoided. The engineer should inform her boss that she has a conflict of interest and that another engineer should make the bearing selection.

Informing authorities of harmful, dangerous, or illegal activities is often called **whistle-blowing.** An engineer who is involved with an organization that is doing these activities has a conflict of interest. He has obligations to protect society, but he also has obligations to his fellow workers and employers. Clearly, the need to protect the public is paramount. However, when performing his public duty, the whistle-blower must be prepared to pay the consequences. He may lose his job or be given a flunky job. He may find it difficult to find new employment because potential employers may be unwilling to hire a whistle-blower. If the engineer has a family, the effects of lost income could be devastating. If he keeps his job, he is likely to be ostracized by his coworkers. Before whistle-blowing, the engineer should try every method possible to persuade the wrongdoers to correct their ways. It takes a lot of strength and courage to do the right thing. Although the consequences of being a whistle-blower may be severe, the consequences of knowing you are unable to do the right thing also can be severe. It is best to avoid this situation as much as possible and work with an honorable company.

No code of ethics can cover every possible ethical situation. Perhaps the simplest guideline is to imagine that you have been selected by the *New York Times* as Engineer of the Year. A reporter follows you around and records all your activities, which are then published daily. Because most of us have an inherent sense of right and wrong, this should lead us to correct ethical behavior.

3.5 RESOURCE ALLOCATION

One of the greatest challenges to society is the proper allocation of resources. When resources are misallocated, it costs lives.

Prisoner's Dilemma

Moral codes exist to enable members of society to cooperate with each other. If each member of society follows the code, behavior is predictable and both the individual and society benefit. However, for personal benefit, an individual can defect from cooperative behavior and take advantage of a code-follower's predictable behavior.

According to ethical egoism, actions are correct that result in self-interest. Thus, should a person seeking maximum self-interest cooperate or defect?

The answer to this question can be determined by playing *games*. A classic game is called the "Prisoner's Dilemma," which goes as follows: The police have just captured two suspects at the scene of a crime. Both suspects are guilty, but the police do not know that. The only way the police can determine guilt is to separate the prisoners and separately interrogate them. While sitting in their separate cells, each prisoner contemplates his strategy. Should he confess or blame the other prisoner for the crime?

If a prisoner confesses, this is cooperative behavior. If both confess, then the guilt can be shared between the two of them and they can plead for a lesser sentence. If one prisoner blames the other for the crime, this defects from cooperative behavior because he is attempting to be set free while having the other prisoner take a harsher sentence.

The situation is not so simple. If both confess, they will share a lesser sentence. If one confesses and the other blames, then the confessor will get a particularly harsh sentence while the blamer is released. However, if each blames the other, the police will know that both are guilty and each will receive a harsh sentence for his crime; their pleadings for a lesser sentence will go unheeded because they both lied to the police. The following *truth table* shows the benefits to each prisoner depending on which actions are taken.

Action		Benefit	
Me	Him	Me	Him
Coop	Coop	3	3
Coop	Defect	0	5
Defect	Coop	5	0
Defect	Defect	1	1

The strategy that maximizes self-interest—a strategy endorsed by ethical egoism—is the subject of **game theory.** In the early 1980s, Robert Axelrod (a political science professor at the University of Michigan) organized a test of computer programs to determine which strategy results in the most self-interest. For a single round of play, defecting always is the best strategy. [The above truth table shows this.] Assuming the other prisoner randomly cooperates or defects,

my average benefit by defecting is 3 (i.e., $(5 + 1)/2$). If I cooperate, my average benefit is only 1.5 (i.e., $(3 + 0)/2$, assuming the other prisoner randomly cooperates or defects.)

In real life, we generally interact with other individuals multiple times. In Axelrod's computer contest, the winning strategy for multiple interactions was "tit-for-tat." This strategy always cooperated in the first round and simply repeated what the opponent did in subsequent rounds, as shown:

1st Round

Me	Him
Coop	Coop
Coop	Defect

⇒

2nd Round

Me
Coop
Defect

The tit-for-tat strategy, which places a premium on cooperative behavior, shows convergence between ethical egoism, utilitarianism, and rights analysis.

In the early 1990s, Martin Nowak (University of Oxford) and Karl Sigmund (Vienna University) showed that when occasional mistakes are made (as occurs in real life), the tit-for-tat strategy degenerates into backbiting. They determined that an alternative strategy was more successful. The "Pavlov" strategy repeats a response if it was successful (i.e., obtains a benefit of 3 or 5) and changes the response if it was unsuccessful (i.e., obtains a benefit of 0 or 1). The truth table for the Pavlov strategy follows.

1st Round

Me	Him	My Benefit
Coop	Coop	3
Coop	Defect	0
Defect	Coop	5
Defect	Defect	1

⇒

2nd Round

Me
Coop
Defect
Defect
Coop

In a population of *generally* cooperative individuals, the Pavlov strategy also leads to cooperation. Thus, in this case, the Pavlov strategy also shows convergence of ethical egoism, utilitarianism, and rights analysis. However, if the population consists of *always* cooperative individuals, Pavlov will exploit these "suckers" by defecting.

In summary, game theory is a tool for evaluating behavioral strategies that conform with ethical egoism. Interestingly, many of the most successful strategies are consistent with the morality of utilitarianism and rights analysis. However, the Pavlov strategy warns that a moral person should be on guard against being exploited as a sucker by less moral individuals.

Adapted from: T. Beardsley, *Scientific American,* October 1993, p. 22.

As an example, consider well-meaning legislation designed to reduce the amount of carcinogens released by the chemical industry. Carcinogens can be reduced to arbitrarily small levels, but always at a cost. If the chemical industry were to spend $1 billion for pollution-control equipment that removes enough carcinogens to save one person's life, would that be a proper allocation of resources? To the person whose life was saved, the answer is yes; but perhaps that $1 billion would be better spent on cancer research or improved prenatal health care. In that case, perhaps hundreds or thousands of lives could be saved.

The big question is, What is the value of a human life? You might reply, "Human life is invaluable; you cannot quantify it with dollars." Although this is a wonderful sentiment, it does not help society to allocate resources. If life has an infinite value, then the chemical industry should install $10 billion of pollution control equipment to save a life, or $100 billion, or $1000 billion. When the chemical industry spends these billions, it loses the opportunity to use those finite resources for other fruitful endeavors. And, of course, the cost is ultimately passed on to the consumer.

In a sense, each of us makes a decision about the value of our own life based upon the car we drive. Every year, about 43,000 people die in traffic accidents; it is the most prevalent cause of accidental death by far. In any year, each of us has about one chance in 7000 of being killed in an automobile accident, assuming we drive an average car in an average manner for an average number of miles. Suppose we are deciding about the purchase of a car we expect to last for 10 years. During the 10-year life of the car, we have one chance in 700 of being killed in an automobile accident if we drive an average car. Fortunately, some cars are built with safety as a high priority (e.g., BMW, Volvo, Lexus, Mercedes Benz, Volkswagen). Imagine we have narrowed our car purchase decision to two models:

Car	Cost	Probability of Fatal Accident During 10-Year Life of Car
Average	$30,000	1/700
Safe	$60,000	1/1400

If 1400 people spend $60,000 for the safe car, they will have spent $84 million and one of them will be dead in 10 years because of a car accident. If 1400 people spend $30,000 for the average car, they will have spent $42 million and two of them will be dead in 10 years because of car accidents. Thus, by spending an extra $42 million, one life can be saved. If you are aware of this safety/cost trade-off and you still select the $30,000 car, you have valued your life at less than $42 million, because you could be that extra person who dies.

Suppose people insisted that a life has infinite value and that engineers must design automobiles that are perfectly safe. The design solution might look something like a tank, cost about $3 million, and have two-gallons-per-mile fuel economy. If we all drove tanks, our roads must be stronger, which means more of us would be killed making stronger roads. Also, the pollution and environmental damage resulting from burning the tremendous amounts of fuel would cost additional lives.

The conclusion is inevitable: Life has a value, and engineers must know what it is so we can make intelligent design decisions. There is no accepted method for determining the value of a human life. Some approaches include determining the amount of lost wages due to an untimely death, determining ransom payments made to kidnappers, or the amount of extra wages workers demand for extra-risky jobs. Valuations of human life range from about $200,000 to $8 million. In his book *Fatal Tradeoffs: Public and Private Responsibilities for Risk,* W. Kip Viscusi argues that the average American assesses his life at $6 million. (Of course, wealthy individuals assess their lives at a higher figure.)

EXAMPLE 3.8

Problem Statement: A highway engineer has $1 million to spend on guardrails for roads. Guardrails cost $13/ft, so 14.5 miles of guardrails can be installed. They have a service life of 20 years. She is considering spending the money on two roads. One is a scenic two-lane road through the mountains that has very steep shoulders. If a car falls off the shoulder, there is certain death for the passenger. The other is a four-lane highway on fairly level ground. If a car drives off the shoulder, there is only a 10% chance of death for the passenger. The scenic two-lane road traffic is 20 cars per day and the four-lane highway traffic is 22,000 cars per day. To help make her decision, she consults *Benefit-Cost Analysis of Road Side Safety Alternatives* (1986, Transportation Research Record 1065). According to this document, at a rate of 20 cars per day on a two-lane road, there will be only 0.01 "encroachments" per mile per year whereas at a rate of 22,000 cars per day on a four-lane road, there will be 3 "encroachments" per mile per year. (An "encroachment" is when a car drives off the road.) Which road should have the guardrails? How much money is spent per life saved? (Assume there is only one occupant per car, and that the presence of the guardrail prevents death from the automobile accident.)

Solution: For the two-lane scenic road,

$$\text{Cost} = \frac{\$1,000,000}{20 \text{ yr}} \times \frac{\text{yr} \cdot \text{mi}}{0.01 \text{ Encroachments}} \times \frac{\text{Encroachment}}{1 \text{ Life saved}} \times \frac{1}{14.5 \text{ mi}} = \frac{\$344,827}{\text{Life saved}}$$

For the four-lane highway,

$$\text{Cost} = \frac{\$1,000,000}{20 \text{ yr}} \times \frac{\text{yr} \cdot \text{mi}}{3 \text{ Encroachments}} \times \frac{\text{Encroachment}}{0.1 \text{ Life saved}} \times \frac{1}{14.5 \text{ mi}} = \frac{\$11,494}{\text{Life saved}}$$

Obviously, the money is better spent on erecting guardrails along the four-lane highway.

Engineers must always keep safety in mind when designing products. The amount of money engineers spend to reduce risk depends on (1) whether the product user voluntarily accepts the risks and (2) the amount of benefit the user derives from the product. For example, driving a car is *much* riskier than living near a nuclear power plant. However, the driver voluntarily accepts these risks and feels that the benefits outweigh the risks. Because radiation release from a power plant accident could be widespread, anyone could be exposed. Therefore, the risk is not voluntary. Also, a person may not feel that electricity derived from nuclear energy has particular value, because it could just as well be derived from other sources. With these considerations, the safety levels demanded in nuclear power plants far exceed those of automobiles.

Risk is something we must deal with every day. Table 3.1 puts many of these risks in perspective.

3.6   CASE STUDIES

We use case studies to illustrate the points made in our discussion. As you read the case studies, place yourself in the role of the engineer and imagine what you would do under the same circumstances.

3.6.1   *Challenger* Explosion

The *Challenger* was one of six space shuttles launched by NASA (Figure 3.2). A space shuttle consists of a reusable delta wing orbiter that contains the main engines, a cargo bay, living quarters, and the cockpit. The main engines are powered by liquid hydrogen and oxygen supplied from an expendable external tank. At liftoff the craft is extremely heavy,

TABLE 3.1
Risk of death for an "average" American[†]

Cause of Death	Probability $\left(\dfrac{\text{deaths}}{\text{person} \cdot \text{year}}\right)$	Cause of Death	Probability $\left(\dfrac{\text{deaths}}{\text{person} \cdot \text{year}}\right)$
From any cause	8585×10^{-6}	Accidents	354×10^{-6}
Disease	7266×10^{-6}	Motor vehicle	173×10^{-6}
Cardiovascular	3622×10^{-6}	Falls	50×10^{-6}
Cancer	2040×10^{-6}	Poison	26×10^{-6}
Diet*	714×10^{-6}	Fire	16×10^{-6}
Tobacco*	612×10^{-6}	Drowning	16×10^{-6}
Sexual behavior*	143×10^{-6}	Choking	13×10^{-6}
Occupation*	82×10^{-6}	Medical complications	10×10^{-6}
Alcohol*	61×10^{-6}	Firearms	6×10^{-6}
Pollution*	41×10^{-6}	Water transport	3×10^{-6}
Industrial products*	$<20 \times 10^{-6}$	Railroad	3×10^{-6}
Food additives*	$<20 \times 10^{-6}$	General aviation	3×10^{-6}
Suicide	120×10^{-6}	Electrocution	2×10^{-6}
Murder (white male)	93×10^{-6}	Commercial aviation	0.8×10^{-6}
Murder (black male)	720×10^{-6}		
Murder (white female)	30×10^{-6}		
Murder (black female)	142×10^{-6}		

*Causes of cancer according to the best estimates published in L. A. Sagan, "Problems in Health Measurements for the Risk Assessor," in *Technological Risk Assessment*. The Hague, Netherlands: Martinus Nijhoff, 1984.
[†]Calculated as the number of occurrences in 1991 divided by the U.S. population.
Adapted from: U.S. Department of Commerce, *Statistical Abstract of the United States,* 1994.

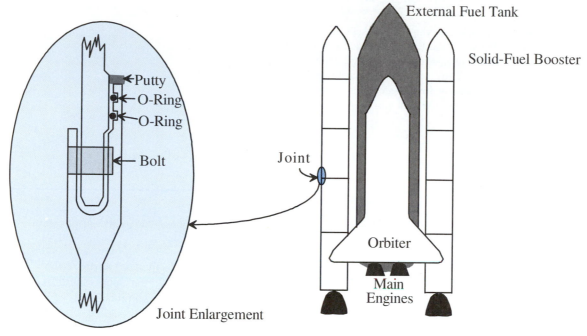

FIGURE 3.2
Challenger space shuttle.

because the filled tanks weigh many millions of pounds. Therefore, two solid-fuel booster rockets are needed during the first 2 minutes of flight. The booster rockets are jettisoned early enough in flight that they do not burn up in the earth's atmosphere and can be recovered for potential reuse.

The booster rockets are extremely large (150 feet long, 12 feet in diameter), so they are difficult to transport. They are assembled from smaller sections that are sealed with vulcanized rubber O-rings and zinc chromide putty (Figure 3.2). The joined sections are then filled with a million pounds of aluminum/potassium chloride/iron oxide propellant. The booster rockets for all three shuttles were manufactured by Morton-Thiokol.

The booster rocket seals had been giving NASA trouble. Inspection of the recovered boosters showed that combustion gases can blow by the O-ring seals, causing them to char. From experience, the engineers knew this problem to be particularly acute during cold weather, when the putty and O-rings are less pliable. Therefore, a program was underway to improve the design.

The space shuttle is a marvelous engineering achievement that brought new capabilities to NASA, such as the ability to repair near-earth satellites. However, it was a program born in controversy. As it was extremely expensive, many other programs were canceled, including heavy-lift expendable rockets. NASA claimed the space shuttle would be an inexpensive "space truck," but it required frequent launches to amortize the development

Risk on the Road

As shown in the figure, automobile driving became safer in the United States between 1950 and 2003. Engineers added safety features to both the automobile (e.g., seat belts, air bags, crumple zones) and roads (break-away signs, crash cushions, guardrails) that contributed to making driving safer, even though the number of cars on the road increased each year.

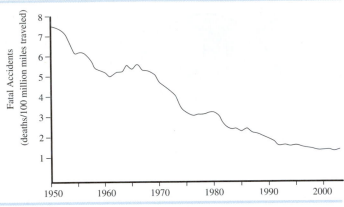

Source: National Highway Traffic Safety Administration (http://www.volpe.dot.gov/infosrc/journal/2005/pdfs/vj05intro.pdf).

costs. Thus, NASA was under pressure to launch frequently, to prove the viability of its enormous investment. In fact, NASA was claiming the space shuttle could generate enough revenue to be self-supporting. A few Reagan administration officials even suggested it be privatized and sold to a commercial airline.

Any large engineering undertaking like the space shuttle should be viewed as an experiment. Although some components can be independently tested (e.g., electronics, engines, pumps), the only way to test the entire system is to launch it. Each launch experiences problems, such as cracked pump blades or some blow-by in the booster seals. Each time a decision is made to proceed with a launch in spite of equipment concerns, it is viewed as a test of the system. Eventually, as the limits of the system are explored, the system will break.

On January 28, 1986, NASA inadvertently tested the temperature limits of the space shuttle *Challenger*. The evening temperature had been below freezing and the morning temperature was 36°F. Prior to all launches, discussions are held between NASA management and engineers about the viability of the upcoming launch. This time, engineers expressed concern that ice at the launch site might damage the orbiter or its fuel tank. Violent waves had forced booster rocket recovery ships to return to the coast. Further, there was engineering concern about the booster rocket O-rings. Roger Boisjoly, a Morton-Thiokol engineer, recommended that no launch should occur below 53°F, based on his previous launch experience. He and his associate, Arnold Thompson, expressed their concerns to Morton-Thiokol and NASA management. Because Morton-Thiokol was negotiating for future booster rocket contracts, they did not want to portray their product as being of poor quality. NASA was interested in an aggressive launch schedule to prove space shuttle economic viability. Thus, the many engineering concerns were overruled by management. The decision was made to launch the space shuttle at 11:38 A.M. After 76 seconds of flight, at 50,000 feet, the booster rocket joints failed, allowing hot escaping gases to ignite the hydrogen fuel. All seven crew members, including Christa MacAuliffe ("teacher in space"), died, as millions watched on television.

Ethical points for discussion:

1. Should the Morton-Thiokol engineers have blown the whistle and announced to the press that NASA management was endangering the lives of the crew members and risking destruction of the space shuttle?

2. No launch is completely safe; space travel is inherently risky. The astronauts accepted this risk when they volunteered for the job. Should they have been informed that the risks were higher for this particular launch?

3. Of the three main moral theories—ethical egoism, utilitarianism, and rights analysis—which moral theory were the Morton-Thiokol and NASA management using when they made the decision to launch over the objections of the engineers? In retrospect, we could judge their action to be wrong. What moral theory or theories are we applying?

4. Is it unfair to place blame on NASA managers? After all, every launch has risks and someone has to make the decision to launch.

5. Should the engineers be faulted for developing an inferior design? Perhaps they should have incorporated heating tape into the joints if low temperatures were known to be a problem.

6. Should an escape mechanism be installed on space shuttles even though it imposes a severe weight penalty and reduces the shuttle payload capability?

Challenger explosion.

Photo courtesy of NASA/NASA Media Services.

3.6.2 Missouri City Television Antenna Collapse

In 1982, a television antenna was erected in Missouri City, Texas. An engineering company designed it, and a separate rigging company erected it. The rigging company approved the detailed engineering drawings and said they could erect the antenna as designed.

The design consists of a three-legged tower constructed from premade sections (Figure 3.3). A movable crane sits at the top of the already-erected sections. It lifts a new section into place, which is then attached to the tower by the riggers. The crane is designed so it can crawl up the tower, allowing it to install the next section at the top. In a sense, the tower is lifted up by its boot straps. All the tower sections are identical except for the top one, which has the microwave basket attached.

The bottom sections of the three-legged tower were successfully erected by the rigging company. The final step was to hoist the top section. When the riggers attempted to attach the cables to attachment points on the top section of the tower, they found that the microwave basket interfered with the cables. They called the engineering company to see if the microwave basket could be removed, allowing the top section to be lifted just like all the others. The microwave basket would be lifted independently and attached after the tower was completely erected. The engineering company refused; their past experience with this

FIGURE 3.3
The Missouri City TV tower disaster.

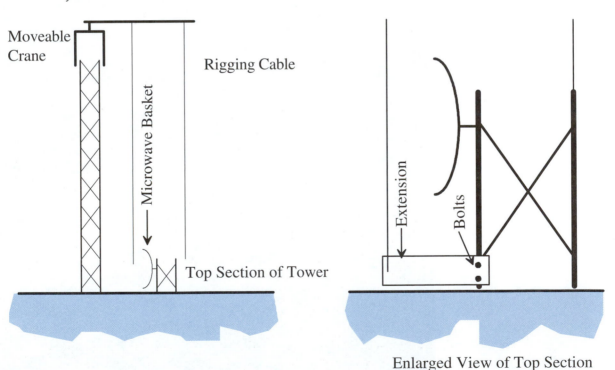

Moveable Crane

Rigging Cable

Microwave Basket

Top Section of Tower

Extension

Bolts

Enlarged View of Top Section
with Added Extension

method was that the microwave basket became damaged, and they were held liable for the damage, so they lost money when they repaired it. The engineers informed the riggers that if they removed the microwave basket, it would void the warranty.

Later, the rigging company contacted the engineering company with a proposal to add an extension to the top section that kept the rigging cables away from the microwave basket. The rigging company did not have in-house engineering expertise and wanted the engineering company to look at a drawing of their proposed modification. Although the engineer wanted to help the riggers, his boss was firmly against the idea. If the engineering company approved the modifications, they would be held liable if anything went wrong. The engineering company reminded the riggers that they had already approved the original drawings; it was the riggers' responsibility to determine how to raise it.

The riggers knew the weight of the top section, so they bought bolts that were rated sufficiently strong to carry the weight. They installed the extension and attached the cables to lift the top section. Six riggers rode with the top section because they were needed to install it once it got to the top. They never made it, because the bolts sheared, causing them all to fall to their deaths. The riggers had neglected to allow for the extra moment arm of the extension, which increased the forces on the bolts by about 12 times.

Ethical points for discussion:

1. Did the engineering company have a faulty design? Should they have anticipated the problem and designed accordingly?
2. Were the riggers operating outside their area of expertise? Should they have hired an engineering consultant to calculate the necessary bolt size?
3. Should the engineering company have been more helpful and less legalistic?

Wreckage from the Missouri City television antenna collapse. The movable crane from the top of the tower is lying in the foreground.

Photo by Lee Lowery, Jr. Texas A&M University/Lee Lowery c/o Civil Engineering.

4. Where did the responsibility of the engineers end and the riggers begin?
5. Of the three major moral theories (ethical egoism, utilitarianism, rights analysis), which theory was the boss of the engineering company using?
6. Elaborate on the legal and moral conflicts faced by the engineer who was contacted by the rigging company.

3.6.3 Kansas City Hyatt Regency Walkway Collapse

In 1976, Crown Center Redevelopment Corporation decided to build a Hyatt Regency Hotel in Kansas City, Missouri. G.C.E. International, Inc. was selected as the structural engineering firm, and Havens Steel Company did the steel fabrication. The building was to have 750 rooms and a large atrium measuring 117 × 145 feet on the floor and 50 feet high. Three suspended walkways through the atrium connected the second, third, and fourth floors. From these walkways, patrons could look down into the atrium, adding to the dramatic effect.

In spring 1978, construction began. In January 1979, Havens contacted Daniel M. Duncan (a G.C.E. engineer) about a proposed design change. The original design specified the suspended walkways for floors two and four to be connected to the ceiling by a single 46-foot threaded rod (Figure 3.4). Havens determined the single rod was simply too long

FIGURE 3.4

Detail of walkway support structure. In the original design, each nut has a load of $\frac{1}{2}F$. In the revised design, the nuts holding the fourth-floor walkway have load F, which is twice their anticipated load.

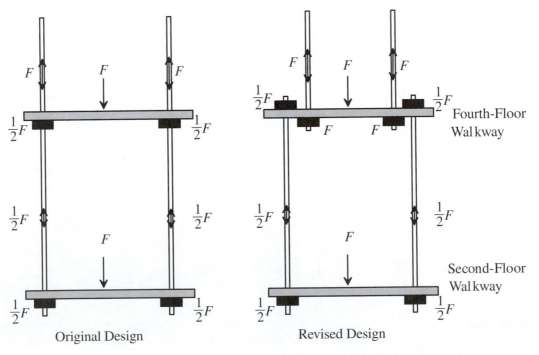

Original Design

Revised Design

to build, so they proposed the threaded rod be separated into two. The design changes were incorporated into shop drawings that received the engineering seal of approval by Jack D. Gillum, president of G.C.E.

In July 1980, the hotel was open for business. One year later, on July 17, 1981, during a tea dance held at the hotel, many party-goers were dancing, standing, and walking on the atrium walkways. The fourth-floor walkway collapsed onto the second-floor walkway, killing 114 and injuring over 200. It was the most deadly structural failure in U.S. history before the collapse of the World Trade Center Twin Towers on September 11, 2001.

In November 1985, Gillum and Duncan were found guilty of gross negligence, misconduct, and unprofessional conduct in the practice of engineering. The collapse brought great scrutiny to their work and revealed many irregularities. Duncan's design did not employ adequate stiffeners at the joints between the threaded rod and walkway. Critical welds near the joints were not specified, and the material selected for the threaded rods was too weak. Duncan failed to review the shop drawings that described the new two-rod support system. Most importantly, Duncan failed to perform engineering calculations of either the original design or the modification. Had he done so, he would have learned that the modification doubled the load on the fourth-floor walkway support bolt. This exceptionally high load caused the joint to fail and led to the walkway collapse. Although Gillum had delegated responsibility to Duncan for performing the walkway design, because Gillum placed his engineering seal of approval on the drawings and had not adequately checked Duncan's work, he was held "vicariously liable" and was subjected to the same penalties as Duncan.

Although Duncan and Gillum lost their Missouri licenses, they are practicing in other states.

Ethical points for discussion:

1. Should these engineers be allowed to practice engineering in other states?
2. Should the engineers be held responsible for a simple error such as this? After all, they had to make thousands of decisions and everyone is entitled to make a mistake.
3. Engineering is a discipline requiring attention to detail. Is it ethical for professors to give students partial credit for wrong answers that are sort of right?
4. In court testimony, G.C.E. claims it was never called by Havens about the proposed design change, yet their seal was affixed to the revised drawings. Does it appear G.C.E. was telling the truth?
5. Should the fabricator be held responsible for overloading the support nut?
6. What can be done to prevent such catastrophes in the future? Should the government review all drawings? If procedural changes are implemented, what is their impact on efficiency and cost?
7. Should the engineers have lost their licenses?
8. Upon further investigation it was found that the atrium roof had collapsed during construction. G.C.E. claims that the owner (Crown Center Redevelopment Corporation) was unwilling to pay for onsite inspection, even though G.C.E. requested it three times. How much responsibility does the owner have to ensure that the engineers are funded well enough to properly perform their jobs?
9. Even the original walkway design was questionable. Were the engineers living up to their obligations to be technically competent?

Wreckage from the walkway collapse of the Kansas City Hyatt Regency.

Photo by Lee Lowery, Jr. Texas A&M University/Lee Lowery c/o Civil Engineering.

10. Because Gillum had delegated responsibility to Duncan to do the design, should Gillum be held liable for Duncan's mistakes?
11. How can competent engineers protect themselves from the impact of incompetent engineers?
12. Should stronger certification laws be enacted to eliminate incompetent engineers?

3.6.4 Hurricane Katrina

New Orleans, Louisiana, is a historic coastal city with a vibrant culture noted for good food, jazz music, and festive celebrations. The most famous section of the city is the French Quarter, much of it built in the 18th and 19th centuries. This and other old portions of the city are about 5 feet above sea level. Newer sections of the city are about 8 feet below sea level and rely upon a system of levees and pumps to keep the land dry. The newer sections were built on silt deposits that subside at a rate of about 2–5 feet per century. Before the levees were constructed, new silt deposits would counteract the subsidence. Unfortunately, levees along the Mississippi River now direct silt into the ocean rather than to the land surrounding New Orleans.

As a coastal city, New Orleans is subjected to periodic hurricanes. Since 1852, it has had seven category 3 (111–130 mph winds), eight category 4 (131–155 mph winds), and two category 5 (>155 mph winds) hurricanes.

On August 29, 2005, Hurricane Katrina hit Louisiana as a strong category 3 hurricane with 125 mph winds. Katrina was one of the deadliest hurricanes in U.S. history, with over 1800 deaths attributed to her wrath, of which almost 1600 occurred in Louisiana alone. It was the costliest hurricane in U.S. history, with over $81 billion in damages. Although it

Jocelyn Augustino/FEMA

affected many communities, the impact on New Orleans was most dramatic—nearly 80% of the city was flooded. Because of poor planning and insufficient transportation, thousands of people remained in the city, many seeking sanctuary in the Superdome sports stadium. Looting was widespread as the police and National Guard troops had difficulty re-establishing social order.

The aftermath of the storm included attempts to assign blame for the disaster to local, state, and federal governments. Because of poor planning, lack of coordination, and delayed responses, blame could be assigned to all levels of government.

The levees were designed to withstand a category 3 hurricane. The levees were overtopped by storm surges. Once water reached the back side of the levee, it carved out soil that supported the structures. Sections of the levees failed in approximately 20 locations. There were intense investigations into the causes of their failure, and the following accusations were made:

- *Government corruption.* The New Orleans Levee Board used monies designated for the levees for other purposes, such as establishing gambling casinos and fiber-optic networks. There were charges that no-bid contracts were issued to contractors with political connections, and that some of the contractors had links to organized crime.

- *Contractor corruption.* The earthen levees were supposed to be constructed from strong fill dirt, but some contractors were accused of cutting corners by using substandard sand, "shell fill," and "swamp muck" dredged from the nearby waterway.

- *Environmentalist lawsuits.* Several environmental groups sued the Army Corps of Engineers to stop a 1996 plan to strengthen levees along the Mississippi River. The purpose of the lawsuit was to allow periodic flooding of the Atchafalaya Basin, which occurred frequently before the levees were constructed, and was considered part of the natural cycle.

- *Insufficient wetlands.* Wetlands can serve as a buffer to dampen the effects of storm surges. (Each 2.7 miles of wetlands removes about 1 foot from a storm surge.) Because of subsidence, wetlands near New Orleans are reduced by about 16,000 acres each year. Prior to the construction of levees, silt would deposit and preserve wetlands. Unfortunately, the levees now channel silt into the ocean, preventing wetland renewal.

- *Inadequate infrastructure.* The water pumps were powered by the electrical grid, which was destroyed by the hurricane. Backup generators should have been installed, which would allow the pumps to continue operating in the event of grid failure.

- *Lack of preparation.* Barges could have been positioned to bring fill material to seal broken levees.

- *Wayward barge.* An unsecured grain barge was suspected to have rammed the levee, causing a breach.

- *Poor maintenance.* Because the land is constantly sinking, it is necessary to add new material to the levees to ensure the same level of protection. This was not done to many sections of the levee.

- *Poor design.* The levees were constructed from many materials, including concrete, soil, and sheet metal pilings. The transition points between different methods of construction were particularly vulnerable.

The American Society of Civil Engineers (ASCE) conducted a detailed study of the levee failure shortly after the disaster and reported to the U.S. Senate Committee on Homeland Security and Governmental Affairs. They determined there were a number of failure mechanisms: scour erosion caused by overtopping, seepage through the walls, soil failure at the foundations, and "piping," an internal erosion caused by water flow. In some locations, the surge was concentrated and funneled by channels in the Mississippi River Gulf Outlet, which caused very high surge levels that overtopped the levees.

Repairing the damage and constructing stronger levees is estimated to cost $14 billion.
Ethical points for discussion:

1. Members of the New Orleans Levee Board are appointed by the governor, and generally are politicians or businesspeople. Should some or all members be required to have an engineering degree?
2. How should society balance the need to preserve natural wetlands with the need to control flooding?
3. Should resources be spent to construct wetlands as buffers?
4. The levee system was designed for a category 3 hurricane. Given that numerous category 4 and 5 hurricanes have hit New Orleans in the past, should funding have been provided to build a stronger levee system?
5. Should a double-wall levee system be built so that a second wall backs up a breach in the first wall? Is the additional expense warranted?
6. Should federal tax money be spent for levees that benefit only the people of New Orleans?
7. Some climate studies suggest that global warming will cause ocean levels to rise and intensify hurricanes because of higher ocean temperatures. Global warming is caused primarily by carbon dioxide released from the combustion of fossil fuels. Does the

entire nation have an obligation to spend money to protect New Orleans from hurricanes and rising ocean levels?

8. Given that the majority of New Orleans is below sea level and that strong hurricanes are likely to hit many times in the future, is it folly to continue to spending money to protect the city?

9. What penalties should be given to contractors who cut corners and use substandard building materials in the levee construction?

10. How should engineers design for unexpected events, such as the grain barge that was suspected of hitting the levee?

11. What is the role of professional societies, such as the ASCE, in protecting society?

3.7   SUMMARY

As professionals, engineers are expected to behave in an ethical manner. Ethical rules are among the interaction rules that govern the relationship between individuals and society; other interaction rules are classified as etiquette, law, and morals. Generally, there is consistency among these various interaction rules, but occasionally, they conflict.

The purpose of interaction rules is to eliminate conflict; however, it is impossible to eliminate all conflicts, so the source of the conflict must be identified. Conflicts can result from moral issues, conceptual issues, application issues, or factual issues. Factual issues are concrete and can generally be settled regardless of a person's upbringing and background. In contrast, moral issues tend to be abstract. Resolution of moral conflicts depends upon a person's upbringing and background.

When attempting to settle moral issues, various moral theories may be employed. Ethical egoism states that actions are moral if you act in your enlightened self-interest. Utilitarianism states that the most moral action is the one that brings the most good to the most people. Rights analysis states that actions are moral if they do not violate the rights of individuals. Often, these three moral theories converge, indicating that one action is clearly correct. However, sometimes they diverge, making it difficult to select a moral action. Each society deals with this problem differently. Some philosophers recommend using utilitarianism unless an individual's rights are seriously impaired.

A major issue confronting society is the proper allocation of resources. Although we all would like to live in a risk-free society, there are insufficient resources to achieve that. Often, engineers must make difficult decisions regarding acceptable levels of risk and safety.

Further Readings

Harris, C. E., Jr., M. S. Pritchard, and M. J. Rabins. *Engineering Ethics: Concepts and Cases,* 3rd ed. Belmont, CA: Wadsworth, 2004.

Johnson, D. G. *Ethical Issues in Engineering.* Englewood Cliffs, NJ: Prentice Hall, 1990.

Martin, M. W., and R. Schinzinger. *Ethics in Engineering,* 4th ed. New York: McGraw-Hill, 2004.

PROBLEMS

3.1 In the following examples, determine whether the source of the conflict is a moral issue, conceptual issue, application issue, or factual issue.

(a) John is a newly hired engineer for the JMT Electronics Company. He is responsible for selecting vendors that supply components for his company's products. One of the vendors invites John to lunch at Chez Pierre Restaurant so they can become better acquainted. When John returns from lunch, his boss is angry with him. He explains that John could be fired for accepting a bribe, because a typical lunch at Chez Pierre costs $50.

(b) In response to the incident described in Part (a), John's boss issues a memo stating that "engineers are not permitted to accept lunch from vendors if the value of the meal exceeds $20." A vendor takes John to another restaurant and spends $19 for the food and leaves a $4 tip. John is concerned that he might be fired for not complying with the memo, because the total spent on him by the vendor was $23, a figure exceeding the policy guidelines. Jane, a fellow engineer, says not to worry, because the food was only $19 and falls within the allowable limit.

(c) During World War II, engineers were important players in the Manhattan Project, the U.S. program that built the first atomic bomb. Mark and Edward were two engineers who debated whether they should participate in the project. Edward argued that they should not become involved because the atomic bomb would mean a level of mass destruction never before unleashed on the world. In his view, humankind is not wise enough to control such power. Mark argued that millions of lives, both U.S. and Japanese, would be sacrificed if conventional weapons were used to end the war. He felt that although the atomic bomb was certainly destructive, many fewer lives would be lost because the war would end much earlier if the atomic bomb were used.

(d) Fred is a quality control engineer in an automotive plant that makes drive shafts on a high-speed lathe requiring specially hardened tool bits. He notices that the automotive drive shafts are no longer meeting specifications on a critical dimension, causing 1000 of them to be rejected. Because his bonus depends on a low rejection rate, Fred is angry at this development. He calls the lathe operator into his office and accuses him of installing the wrong tool bit in the high-speed lathe. The lathe operator swears that he used the right tool bit.

(e) Larry is an aerospace engineer employed by Boeing to design planes. He is a member of the Quaker religion, which is famous for nonviolence. (During wars, Quakers often refuse to serve as fighting soldiers, but will serve as medics.) Larry was hired by Boeing to design passenger airplanes; however, his boss has recently reassigned him to design military fighters. Larry must decide whether to accept the new assignment, or quit and find a new job.

(f) Sally is a mechanical engineer employed by General Motors to design automotive gas tanks. According to U.S. government safety standards, the automobile must survive a moderate impact with no chance of the gas tank catching fire. In recent tests, cars that crashed at 35 mph had no fires, whereas 20% of cars that crashed at 45 mph had fires. She is considering whether to redesign the gas tank.

(g) A new government law requires that the lead content of drinking water be less than 1.0 ppb (parts per billion). Melissa is a safety engineer who has tested her company's drinking water by two methods. Method A gives a reading of 0.85 ppb, whereas Method B gives a reading of 1.23 ppb. She must fill out a government report describing the quality of her company's water. If the lead content exceeds 1.0 ppb, her company will be fined. She is contemplating whether to report the results from Method A or Method B.

3.2 The environmental organization Greenpeace wants to stop "toxic colonialism," in which developed nations send their hazardous waste to less-developed countries with lax environmental laws. The United Nations has proposed regulations titled "The Basel Convention on the Control of Transboundary Movements of Hazardous Wastes and Their Disposal." The regulations seek to ban the export of hazardous waste from developed to lesser developed countries. In 2006, 168 nations were signatories to the agreement.

Mary is an engineer who works for Copper Recyclers, Inc. They ship bales of used copper wire from the United States to Mexico, where they have a processing plant. Because the copper is used, it is contaminated with small amounts of tin-lead solder. She is concerned that the Basel Convention will affect her company's business because of the lead content in their bales. She must make a recommendation to her boss whether to relocate the processing plant in the United States.

From this scenario, identify a moral issue, a conceptual issue, an application issue, and a factual issue.

3.3 In the 1990s, Volvo introduced an automobile to the U.S. market in which the headlights were on whenever the engine was on; thus, the lights were on even during the day. In Sweden, the law required all motorists to drive with their headlights on even during the day. Since enacting this law, Swedish motorist deaths had been reduced by 11%. Because of the extra energy used by the headlights and the need for more frequent replacement of the headlights, this practice added approximately $0.002/mi. In the 1990s, the U.S. government had to decide if they wished to make this feature standard on all U.S. automobiles. At the time, there were 180 million automobiles and 250 million people in the United States. Each vehicle was driven about 15,000 miles each year. You were hired to help make the decision, so perform the following calculations:

(a) How many lives would be saved?

(b) What is the cost of saving each life?

(c) Assuming a human life is valued at $7 million, is this a legitimate expense?

3.4 A retired engineer is contemplating whether to fly or drive to a destination that is 1000 miles away. He has heard that flying is safer than driving and decides to calculate the probability of being killed while traveling.

(a) Calculate the probability of the retired engineer being killed on a 1000-mile one-way trip taken by car and by plane. Use the following assumptions: There are about 180 million automobiles in the United States and 300 million people. Each vehicle is driven about 15,000 miles per year. An average American travels 1800 miles by plane each year. The engineer is in good health and has driving skills comparable to the average American.

(b) The direct cost of driving (fuel, oil, wear-and-tear) is about $0.38/mi. (*Note:* This cost does not include fixed costs such as insurance, registration fees, automobile depreciation, and interest. These direct costs must be paid regardless of whether he drives or not; therefore, they will not be charged to the trip. If both fixed and direct costs are included, the typical cost of driving an automobile is about $0.68/mi.) A round-trip airplane ticket for the 1000-mile trip costs $700. By flying rather than driving, how much money is spent to save a life? The engineer places no value on his time because he is retired, but he does place a $7 million value on his life. Is the extra cost of the airplane ticket worth it to the retired engineer?

(c) The above calculation shows flying to be much safer than driving on a per-mile basis. The retired engineer spends two hours in the car getting to the airport and two hours on the plane. Is it valid for him to feel much safer during the airplane trip and much less safe during the automobile trip? In making your calculation, use the following assumptions: Automobiles average about 35 mph and airplanes average about 500 mph.

3.5 Prepare truth tables for the Prisoner's Dilemma game using the following strategies. Play four rounds using each strategy. Report your "benefit" after each round. Compute your average benefit for each strategy after the four rounds are completed. Be sure to evaluate all possible initial conditions.

(Problem	My Strategy	His Strategy
a.	Random*	Random*
b.	Random	Always cooperate
c.	Random	Always defect
d.	Tit-for-tat	Random
e.	Tit-for-tat	Always cooperate
f.	Tit-for-tat	Always defect
g.	Tit-for-tat	Tit-for-tat
h.	Pavlov	Random
i.	Pavlov	Always cooperate
j.	Pavlov	Always defect
k.	Pavlov	Tit-for-tat
l.	Pavlov	Pavlov

*Random means that in an unpredictable manner, about half the time the prisoner cooperates and about half the time the prisoner defects.

3.6 Write a computer program that allows you to play against the computer. Program the computer to follow the random, tit-for-tat, or Pavlov strategy.

3.7 Write a computer program that calculates the average benefit to you after playing 100 rounds of one strategy listed in Problem 3.5.

3.8 Write a computer program that calculates the average benefit to you after playing 100 rounds of one strategy listed in Problem 3.5. Include in your program a feature that allows for occasional mistakes. For example, rather than always cooperating, perhaps the other prisoner cooperates 90% of the time and defects 10% of the time.

Glossary

$\sum$ The Greek symbol for sum.

application issue The lack of clarity as to whether a particular act violates a law, rule, or policy.

conceptual issue The morality of an action is agreed upon, but there is uncertainty about how it should be codified into clearly defined law, rule, or policy.

conflict of interest A situation in which an engineer's loyalties and obligations may be compromised because of self-interest or other loyalties and obligations.

convergence To tend toward a common solution.

diverge To differ.

eminent domain The right of the government to take possession of private property for public use.

engineering ethics The set of behavioral standards that all engineers are expected to follow.

ethical egoism A moral theory stating that an act is moral provided you act in your enlightened self-interest.

ethics The general and abstract concepts of right and wrong behavior culled from philosophy, theology, and professional societies.

etiquette The codes of behavior and courtesy.

factual issue Uncertainty about morally relevant facts.

game theory A tool for evaluating optimal behavior strategies.

happiness objective function $\sum$ (benefit) (importance) $- \sum$ (harm) (importance)

interaction rules Expected sets of behavior (etiquette, law, morals, and ethics) among the engineer, other individuals, and society as a whole.

law The system of rules established by authority, society, or custom.

legal rights The "just claims" given to all humans within a government's jurisdiction.

moral issue An issue that can be resolved only by making a moral decision.

moral rights The "just claims" that belong to all humans, regardless of whether these rights are recognized by government.

moral theories A framework for making moral and ethical decisions.

morals The accepted standards of right and wrong that are usually applied to personal behavior.

rights analysis A moral theory that equally respects each human being.

utilitarianism A moral theory that seeks to create the most good for the most people.

vulcanized To improve the strength and resiliency of a polymer such as rubber or plastic by combining it with sulfur (or some other type of agent).

whistle-blowing The act of informing authorities of harmful, dangerous, or illegal activities.

CHAPTER 4

Problem Solving

Engineers are problem solvers; employers hire them specifically for their problem-solving skills. As essential as problem solving is, it is impossible to teach a specific approach that will always lead to a solution. Although engineers use science to solve problems, this skill is more art than science. The only way to learn problem solving is to do it; thus, your engineering education will require that you solve literally thousands of homework problems.

In the modern world, computers are often used to aid problem solving. The novice student may think that the computer is actually solving the problem, but this is untrue. Only a human can solve problems; the computer is merely a tool.

4.1 TYPES OF PROBLEMS

A *problem* is a situation, faced by an individual or a group of individuals, for which there is no obvious solution. There are many types of problems we confront:

- *Research problems* require that a hypothesis be proved or disproved. A scientist may hypothesize that CFCs (chlorofluorocarbons) are destroying the earth's ozone layer. The problem is to design an experiment that proves or disproves the hypothesis. If you were confronted with this research problem, how would you approach it?
- *Knowledge problems* occur when a you encounter a situation that you do not understand. A chemical engineer may notice that the chemical plant produces more product when it rains. The cause is not immediately obvious, but further investigation might reveal that heat exchangers are cooled by the rain and hence have more capacity.
- *Troubleshooting problems* occur when equipment behaves in unexpected or improper ways. An electrical engineer may notice that an amplifier has a 60-cycle hum whenever the fluorescent lights are turned on. To solve this problem, she determines that extra shielding is required to isolate the electronics from the 60-cycle radiation emitted by the lights.
- *Mathematics problems* are frequently encountered by engineers, whose general approach is to describe physical phenomena with mathematical models. If a physical phenomenon can be described accurately by a mathematical model, the engineer unleashes the extraordinary power of mathematics, with its rigorously proved theorems and algorithms, to help solve the problem.

- *Resource problems* are always encountered in the real world. It seems there is never enough time, money, people, or equipment to accomplish the task. Engineers who can get the job done in spite of resource limitations are highly prized and well rewarded.
- *Social problems* can impact engineers in many ways. A factory may be located where there is a shortage of skilled labor because the local schools are of poor quality. In this environment, an engineer running a training program for factory workers must design the program to accommodate the low reading abilities of the trainees.
- *Design problems* are the heart of engineering. To solve them requires creativity, teamwork, and broad knowledge. A design problem must be properly posed. If your boss said, "Design a new car," you would not know whether to design an economy car, a luxury car, or a sport/utility vehicle. A well-posed design problem must include the ultimate objectives of the design project. If the boss said, "Design a car that goes from 0 to 60 miles per hour in 6.0 seconds, gets 50 miles-per-gallon fuel economy, costs less than $20,000, meets government pollution standards, and appeals to aesthetic tastes," then you could begin the project—even though it has difficult objectives.

4.2 PROBLEM-SOLVING APPROACH

The approach to solving an engineering problem should proceed in an orderly, stepwise fashion. The early steps are qualitative and general, whereas the later steps are more quantitative and specific. The elements of problem solving can be described as follows:

1. *Problem identification* is the first step toward solving a problem. For students, this step is done for them when the professor selects the homework problems. In the real engineering world, this step is often performed by a manager or creative engineer.

   As an example, the management of an automotive firm may be painfully aware that the firm is losing market share. They challenge the engineering staff to design a revolutionary automobile to gain back lost sales.

2. *Synthesis* is a creative step in which parts are integrated together to form a whole.

   For example, the engineers may determine that they can meet the design objectives for the new car (high fuel economy and rapid acceleration) by combining a highly efficient engine with a sleek, aerodynamic body.

3. *Analysis* is the step where the whole is dissected into pieces. Most of your formal engineering education will focus on this step. A key aspect of analysis is to translate the physical problem into a mathematical model. Analysis employs logic to distinguish truth from opinion, detect errors, make correct conclusions from evidence, select relevant information, identify gaps in information, and identify the relationship among parts.

   For example, the engineers may compare the drag of a number of different body types and determine if the engine can fit under the hood of each body.

4. *Application* is a process whereby appropriate information is identified for the problem at hand.

   For example, the engineers determine that a key question is to find the required force needed to propel the automobile at 60 mph at sea level, knowing the car has a projected frontal area of 19 ft^2 and a drag coefficient of 0.25.

5. *Comprehension* is the step in which the proper theory and data are used to actually solve the problem.

For example, the engineers determine that the drag force F on the automobile may be calculated using the formula

$$F = \tfrac{1}{2}C_d\rho Av^2$$

where C_d is the drag coefficient (dimensionless), ρ is the air density (kg/m^3), A is the projected frontal area (m^2), v is the automobile velocity (m/s), and F is the drag force (N). From the data, the force required to overcome air drag is

$$F = \tfrac{1}{2}(0.25)\left(1.18 \ \frac{\text{kg}}{\text{m}^3}\right)\left[19 \ \text{ft}^2 \times \left(\frac{\text{m}}{3.281 \ \text{ft}}\right)^2\right]$$

$$\times \left(60 \ \frac{\text{mi}}{\text{h}} \times \frac{\text{h}}{3600 \ \text{s}} \times \frac{5280 \ \text{ft}}{\text{mi}} \times \frac{\text{m}}{3.281 \ \text{ft}}\right)^2 \times \frac{\text{N}}{\dfrac{\text{kg·m}}{\text{s}^2}}$$

$$= 190 \ \text{N} \times \frac{\text{lb}_\text{f}}{4.448 \ \text{N}} = 42 \ \text{lb}_\text{f}$$

The required force to overcome air drag is 190 newtons (in the metric system) or 42 pounds-force (in the American Engineering System). (*Note:* The above calculation involves many **conversion factors** from Appendix A. If you are uncomfortable with the conversions, do not worry; they will be discussed in more detail later in the book. You may wish to look at "A Word About Units" to review some basics regarding units.)

A Word about Units

Undoubtedly, you have been exposed to units of measure in high school. Here, our purpose is to refresh your memory. You are certainly familiar with the formula relating distance d, speed s, and time t,

$$d = st$$

Suppose your speed is 60 miles per hour and your trip takes 2 hours. We can calculate the distance traveled as

$$d = \frac{60 \ \text{miles}}{\text{hour}} \times 2 \ \text{hour} = 120 \ \text{miles}$$

The hours cancel, leaving units of miles. Some students prefer to show their calculations as follows:

$$d = \frac{60 \ \text{miles}}{\text{hour}} \left| \frac{2 \ \text{hour}}{} \right. = 120 \ \text{miles}$$

Either approach is correct as long as you are careful with your units.

If you wish to express the distance in kilometers, then a conversion factor is required. There is about 0.6 mile in a kilometer. Use this relationship to convert miles to kilometers:

$$d = 120 \ \text{miles} \times \frac{\text{kilometer}}{0.6 \ \text{mile}} = 200 \ \text{kilometers}$$

When a unit is raised to a power, the conversion factor also must be raised to a power. For example, if a large land area were 120 square miles, then the area would be converted to square kilometers as follows:

$$A = 120 \ \text{miles}^2 \times \left(\frac{\text{kilometer}}{0.6 \ \text{mile}}\right)^2 = 333 \ \text{kilometers}^2$$

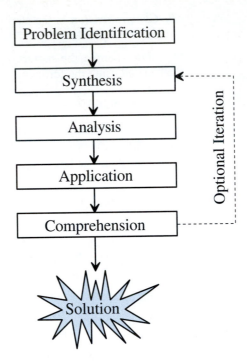

FIGURE 4.1
Problem-solving approach.

Although we would like to believe that these five steps can be followed in a linear sequence that always leads us to the correct solution, such is not always the case. Often, problem solving is an **iterative procedure,** meaning the sequence must be repeated because information learned at the end of the sequence influences decisions early in the sequence (Figure 4.1). For example, if the engineers determine that the force calculated in Step 5 is too high, they would have to return to earlier steps and try again.

4.3   PROBLEM-SOLVING SKILLS

Problem solving is a process in which an individual or a team applies knowledge, skills, and understanding to achieve a desired outcome in an unfamiliar situation. The solution is constrained by physical, legal, and economic laws as well as by public opinion.

To become a good problem solver, the engineer must have the following:

- Knowledge (first acquired in school, but later on the job).
- Experience to wisely apply knowledge.
- Learning skills to acquire new knowledge.
- Motivation to follow through on tough problems.
- Communication and leadership skills to coordinate activities within a team.

Table 4.1 compares skilled and novice problem solvers. Among the most important capabilities of a skilled problem solver is **reductionism,** the ability to logically break a problem into pieces. (Question: How do you eat an elephant? Answer: One bite at a time.)

TABLE 4.1
Comparison of skilled and novice problem solvers

Characteristic	Skilled Problem Solver	Novice Problem Solver
Approach	Motivated and persistent Logical Confident Careful	Easily discouraged Not logical Lacks confidence Careless
Knowledge	Understands problem requirements Rereads problem Understands facts and principles	Does not understand problem requirements Relies on a single reading Cannot identify facts and principles
Attack	Breaks the problem into pieces* Understands the problem before starting	Attacks the problem all at once Tries to calculate the answer right away
Logic	Uses basic principles Works logically from step to step	Uses intuition and guesses Jumps around randomly
Analysis	Organized Thinks carefully and thoroughly Clearly defines terms Careful about relationships and meaning of terms	Disorganized Hopes the answer will come Uncertain about the meaning of symbols Jumps to unfounded conclusions about the meanings of terms
Perspective	Has a feel for the correct magnitude of answers Understands the differences between important and unimportant issues Uses rule of thumb to estimate the answer	Uncritically believes the answers produced by the calculator or computer Cannot differentiate between important and unimportant issues Cannot estimate the answer

* Very important
Adapted from: H. S. Fogler, "The Design of a Course in Problem-Solving," in *Problem Solving,* AIChE
Symposium Series, vol. 79, no. 228, 1983.

Reductionism contrasts with *synthesis,* the creative process of putting pieces together. If your problem were to design an airplane, you would use reductionism to design subsystems (engines, landing gear, electronic controls, etc.) and synthesis to combine the pieces together.

4.4   TECHNIQUES FOR ERROR-FREE PROBLEM SOLVING

All students hope to do error-free problem solving, because it assures them of excellent grades on homework and exams. Further, error-free calculations will be required when the student enters the engineering workforce. The truth is that engineers can never be certain that their answers are correct. A civil engineer who goes through elaborate calculations to design a bridge cannot be certain of her calculations until a heavy load is placed on the bridge and the deflection agrees with her calculations. Even then there is uncertainty, because there may be errors in the calculations that tend to cancel each other out.

Although we can never be certain our answer is correct, we can increase the probability of calculating a correct answer using the following procedure:

Given:
1. Always draw a picture of the physical situation.
2. State any assumptions.
3. Indicate all given properties on the diagram *with their units.*

Find:
4. Label unknown quantities with a question mark.

Relationships:
5. From the text, write the *main equation* that contains the desired quantity. (If necessary, you might have to derive the appropriate equation.)
6. Algebraically manipulate the equation to isolate the desired quantity.
7. Write *subordinate equations* for the unknown quantities in the main equation. Indent to indicate that the equation is subordinate. You may need to go through several levels of subordinate equations before all the quantities in the main equation are known.

Solution:
8. After all algebraic manipulations and substitutions are made, insert numerical values *with their units*.
9. Ensure that units cancel appropriately. Check one last time for a sign error.
10. Compute the answer.
11. Clearly mark the final answer. *Indicate units.*
12. Check that the final answer makes physical sense!
13. Ensure that all questions have been answered.

Notice that the first step in solving an engineering problem is to draw a picture. We cannot overstate the importance of graphics in engineering. Solving most engineering problems requires good visualization skills. Further, communicating your solution requires the ability to draw.

The above procedure is a process of working backward. You start from the end with the unknown, desired quantity and work backward using the given information. Not every problem fits into the paradigm described above, but if you use it as a guide, it will increase the likelihood of obtaining a correct answer. Notice that units are emphasized. Most calculation errors result from a mistake with units.

It goes without saying that the solution should be as neat as possible so it can be easily checked by another person (a co-worker or boss). Calculations should be performed in pencil so changes can be made easily. It is recommended that you use a mechanical pencil because it does not need to be sharpened. A fine (0.5-mm diameter), medium-hardness lead works best. Any erasures need to be complete.

Most engineering professors prefer that students use **engineering analysis paper,** because it has a light grid pattern that can be used for graphing or drafting. It also has headings and margins that allow the calculations to be labeled and documented. Use one side of the paper only. If at all possible, complete the solution on a single page; this allows for easy checking. Sample Problems 1, 2, and 3 present examples of well-solved homework problems.

1.

$$m_{WOOD} = m_{WATER} \quad (ARCHIMEDES \ PRINCIPLE)$$

$$m_{WOOD} = \rho_{WOOD} \ V_{WOOD}$$

$$V_{WOOD} = \pi r^2 L$$

$$m_{WOOD} = \rho_{WOOD} \ \pi r^2 L$$

$$m_{WATER} = \rho_{WATER} \ V_{WATER}$$

$$V_{WATER} = \pi r^2 L_2$$

$$L = L_1 + L_2$$

$$L_2 = L - L_1$$

$$V_{WATER} = \pi r^2 (L - L_1)$$

$$m_{WATER} = \rho_{WATER} \ \pi r^2 (L - L_1)$$

$$\rho_{WOOD} \ \pi r^2 L = \rho_{WATER} \ \pi r^2 (L - L_1)$$

$$\rho_{WOOD} \ L = \rho_{WATER} \ (L - L_1) = \rho_{WATER} \ L - \rho_{WATER} \ L_1$$

$$\rho_{WATER} \ L_1 = \rho_{WATER} \ L - \rho_{WOOD} \ L = L (\rho_{WATER} - \rho_{WOOD})$$

$$L_1 = L \ \frac{\rho_{WATER} - \rho_{WOOD}}{\rho_{WATER}} = 40.0 \ cm \ \frac{1.00 \ g/cm^3 - 0.600 \ g/cm^3}{1.00 \ g/cm^3}$$

$$\boxed{L_1 = 16.0 \ cm}$$

Sample Problem 1:

Archimedes' principle states that the total mass of a floating object equals the mass of the fluid displaced by the object. A 40.0-cm log is floating vertically in the water. Determine the length of the log that extends above the water line. The water density is 1.00 g/cm³ and the wood density is 0.600 g/cm³.

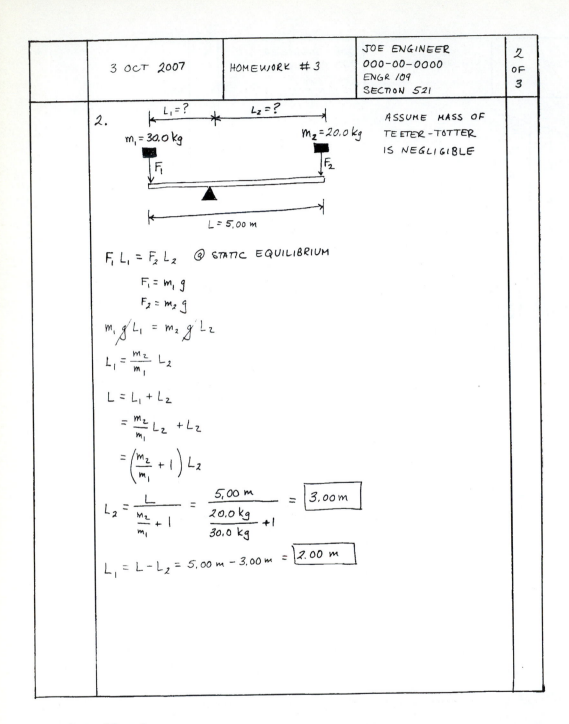

2.

$L_1 = ?$ $L_2 = ?$

$m_1 = 30.0 \text{ kg}$ $m_2 = 20.0 \text{ kg}$ ASSUME MASS OF TEETER-TOTTER IS NEGLIGIBLE

F_1 F_2

$L = 5.00 \text{ m}$

$F_1 L_1 = F_2 L_2$ @ STATIC EQUILIBRIUM

$$F_1 = m_1 g$$

$$F_2 = m_2 g$$

$$m_1 g L_1 = m_2 g L_2$$

$$L_1 = \frac{m_2}{m_1} L_2$$

$$L = L_1 + L_2$$

$$= \frac{m_2}{m_1} L_2 + L_2$$

$$= \left(\frac{m_2}{m_1} + 1 \right) L_2$$

$$L_2 = \frac{L}{\frac{m_2}{m_1} + 1} = \frac{5.00 \text{ m}}{\frac{20.0 \text{ kg}}{30.0 \text{ kg}} + 1} = \boxed{3.00 \text{ m}}$$

$$L_1 = L - L_2 = 5.00 \text{ m} - 3.00 \text{ m} = \boxed{2.00 \text{ m}}$$

Sample Problem 2:

An object is in static equilibrium when all the moments balance. (A moment is a force exerted at a distance from a fulcrum point.) A 30.0-kg child and a 20.0-kg child sit on a 5.00-m long teeter-totter. Where should the fulcrum be placed so the two children balance?

3.

$V = 1.5\,V$

$i = ?$

$R_1 = 5.00\,\Omega$

$R_2 = 10.0\,\Omega$

$R_3 = 15.0\,\Omega$

$$V = iR$$

$$i = \frac{V}{R}$$

$$\frac{1}{R} = \frac{1}{R_1} + \frac{1}{R_2} + \frac{1}{R_3}$$

$$R = \left[\frac{1}{R_1} + \frac{1}{R_2} + \frac{1}{R_3}\right]^{-1}$$

$$= \left[\frac{1}{5.00\,\Omega} + \frac{1}{10.0\,\Omega} + \frac{1}{15.0\,\Omega}\right]^{-1}$$

$$= 2.73\,\Omega$$

$$i = \frac{1.5\,V}{2.73\,\Omega} \times \frac{\Omega}{V/A}$$

$$\boxed{i = 0.55\,A}$$

Sample Problem 3:

The voltage drop through a circuit is equal to the current times the total resistance of the circuit. When resistances are placed in parallel, the inverse of each resistance sums to the inverse of the total resistance. Three resistors (5.00 Ω, 10.0 Ω, and 15.0 Ω) are placed in parallel. How much current flows from a 1.5-V battery?

4.5   ESTIMATING

The final step in solving a problem is to check the answer. One can accomplish this by working the problem using a completely different method, but often there is not time for this. Instead, a valuable approach is to estimate the answer.

The ability to estimate comes with experience. After many years of working similar problems, an engineer can "feel" if the answer is in the right ballpark. As a student, you do not have experience, so estimating may be difficult for you; however, your inexperience is not an excuse for failing to learn how. If you cultivate your estimating abilities, they will serve you well in your engineering career. It has been our experience that much business is conducted over lunch, using napkins for calculations and drawings. An engineer who has the ability to estimate will impress both clients and bosses.

A question we often hear from students is, How do I make reasonable assumptions during the estimation process? There is no quick and simple answer. One of the reasons that "has broad interests" and "collects obscure information" are included among the traits of a creative engineer (Chapter 1) is that these traits will supply you with raw data and cross-checks for reasonableness when making estimations. With practice and experience you will become better. Notice how much your skills improve with the few examples you do in the classroom and for homework. After working in a particular engineering field for a while, many estimations will become second nature.

To get you started as an expert estimator, we have prepared Table 4.2. It lists relationships we find to be very useful during our business lunches. Use this table as a starting point; as you mature in your own engineering discipline, you will undoubtedly learn other relationships and rules of thumb that will serve you well.

The following examples describe five approaches to estimation.

EXAMPLE 4.1   Simplify the Geometry

Problem Statement: Estimate the surface area of an average-sized man.

Solution: Approximate the human as spheres and cylinders.

$$A = A_{\text{head}} + A_{\text{torso}} + 2A_{\text{leg}} + 2A_{\text{arm}}$$

$$= 4\pi r_{\text{head}}^2 + (2\pi rL)_{\text{torso}} + 2(2\pi rL)_{\text{leg}} + 2(2\pi rL)_{\text{arm}}$$

$$= 4\pi(3.75 \text{ in})^2 + [2\pi(5 \text{ in})25 \text{ in}] + 2[2\pi(3 \text{ in})32 \text{ in}] + 2[2\pi(1.5 \text{ in})24 \text{ in}]$$

$$= 2620 \text{ in}^2 \times \left(\frac{2.54 \text{ cm}}{\text{in}}\right)^2 \times \left(\frac{\text{m}}{100 \text{ cm}}\right)^2 = 1.69 \text{ m}^2$$

This is very close to the accepted value of 1.7 m^2.

TABLE 4.2
Useful relationships for estimation purposes

Conversion Factors

1 ft = 12 in	1 atm = 760 mm Hg
1 in = 2.54 cm	1 atm $\approx$ 34 ft H_2O
1 mi = 5280 ft	1 min = 60 s
1 km $\approx$ 0.6 mi	1 h = 60 min
1 ft^3 = 7.48 gal	1 d = 24 h
1 gal = 3.78 L	1 year = $365\frac{1}{4}$ d
1 bbl = 42 gal	1 Btu $\approx$ 1000 J
1 mi^2 = 640 acre	1 Btu $\cong$ $\frac{1}{4}$ kcal
1 kg $\approx$ 2.2 lb_m	1 hp = 550 ft·lb_f/s
1 atm = 14.7 psi	1 hp $\approx$ 0.75 kW
1 atm $\approx$ 1 $\times$ 10^5 N/m^2	

Temperature Conversions

[K] = [°C] + 273.15   [°R] = [°F] + 459.67
[°F] = 1.8 [°C] + 32   [°C] = ([°F] − 32)/1.8

Ideal Gas

Standard temperature = 0°C
Standard pressure = 1 atm
Molar volume (@ STP) = 22.4 L/gmole
Molar volume (@ STP) = 359 ft^3/lbmole

Molecular Weights

H = 1	N = 14
C = 12	O = 16
Air = 29 (21 mol % O_2, 79 mol % N_2)	

Geometry Formulas

Circle area = πr^2
Circle circumference = $2\pi r$
Cylinder volume = $\pi r^2 L$
Cylinder area (without end caps) = $2\pi r L$
Sphere area = $4\pi r^2$
Sphere volume = $\frac{4}{3}\pi r^3$
Triangle area = $\frac{1}{2}$ (base)(height)

Temperature References

Absolute zero = 0 K
He boiling point = 4 K
N_2 boiling point = 77 K
CO_2 sublimation point = 195 K = −78°C
Mercury melting point = −39°C
Freezer temperature = −20°C
H_2O freezing point = 0°C
Refrigerator temperature = 4°C
Room temperature = 20°C to 25°C
Body temperature = 37°C
H_2O boiling point = 100°C
Lead melting point = 327°C
Aluminum melting point = 660°C
Iron melting point = 1540°C
Flame temperature (air oxidant) = 2100°C
Flame temperature (pure oxygen) = 3300°C
Carbon melting point = 3700°C

Physical Constants

Speed of light $\approx$ 3 $\times$ 10^8 m/s
Speed of sound (air, 20°C, 1 atm) $\approx$ 770 mph
Avogadro's number $\approx$ 6 $\times$ 10^{23}
Acceleration from gravity $\approx$ 9.8 m/s^2 $\approx$ 32.2 ft/s^2

Mathematical Constants

$e \approx$ 2.718 $\pi \approx$ 3.14159

Water Properties

Density = 1 g/cm^3 = 62.4 lb_m/ft^3 = 8.34 lb_m/gal
Latent heat of vaporization $\approx$ 1000 Btu/lb_m
Latent heat of fusion $\approx$ 140 Btu/lb_m
Heat capacity = 1 Btu/(lb_m · °F) = 1 kcal/(kg · °C)
Melting point = 0°C
Boiling point = 100°C

Densities

Air (0°C, 1 atm) = 1.3 g/L	Aluminum = 2.6 g/cm^3
Wood $\approx$ 0.5 g/cm^3	Steel = 7.9 g/cm^3
Gasoline = 0.67 g/cm^3	Lead = 11.3 g/cm^3
Oil = 0.88 g/cm^3	Mercury = 13.6 g/cm^3
Concrete = 2.3 g/cm^3	

Earth

Circumference $\approx$ 40,000 km
Mass $\approx$ 6 $\times$ 10^{24} kg

People

Male	Female
Avg mass = 191 lb_m	Avg mass = 166 lb_m
Avg height = 69 in	Avg height = 64 in

Sustained work output of man = 0.1 hp

Energy

Natural gas: 1000 standard ft^3 $\approx$ 10^6 Btu
Crude oil: 1 bbl $\approx$ 6 $\times$ 10^6 Btu
Gasoline: 1 gal $\approx$ 125,000 Btu
Coal: 1 lb_m $\approx$ 10,000 Btu

Ultimate Strength of Materials

Steel = 60,000 to 125,000 lb_f/in^2
Aluminum = 11,000 to 80,000 lb_f/in^2
Concrete = 2000 to 5000 lb_f/in^2

Automobiles

Typical mass = 2000 to 4000 lb_m
Typical fuel usage = 600 gal/year
Typical fuel economy = 20 to 30 mi/gal
Typical power = 50 to 500 hp
Engine speed = 1000 to 5000 rpm

Typical Appliance Energy Usage

Fluorescent lamp = 40 W
Incandescent lamp = 50 to 100 W
Refrigerator-freezer = 500 W
Window air conditioner = 1500 W
Heat pump = 12,000 W
Cooking range = 12,000 W
Avg continuous home electricity usage $\approx$ 1 kW

EXAMPLE 4.2   Use Analogies

Problem Statement: Estimate the volume of an average-sized man.

Solution: Assume the density of a man is 0.95 that of water. (People are mostly water, but they do float slightly when swimming, so the density must be slightly less than water.)

$$V = \frac{m}{\rho} = \frac{191 \text{ lb}_m}{0.95(1.0 \frac{g}{cm^3})} \times \frac{kg}{2.2 \text{ lb}_m} \times \frac{1000 \text{ g}}{kg} \times \left(\frac{m}{100 \text{ cm}}\right)^3 = 0.091 \text{ m}^3$$

EXAMPLE 4.3   Scale Up from One to Many

Problem Statement: How many bed pillows can fit in the back of a tractor trailer?

Solution: One pillow measures 3 in thick by 16 in wide by 21 in long. The cargo bed of a tractor trailer measures roughly 8 ft wide by 10 ft tall by 35 ft long.

$$\text{Number of pillows} = \frac{V_{truck}}{V_{pillow}} = \frac{(8 \text{ ft})(10 \text{ ft})(35 \text{ ft})}{(3 \text{ in})(16 \text{ in})(21 \text{ in})} \times \left(\frac{12 \text{ in}}{\text{ft}}\right)^3 = 4800$$

This calculation neglects the compression of the pillows as they stack on top of each other (which would increase the number carried). It also neglects the volume occupied by packing materials (which would decrease the number carried). One can hope that the errors due to the simplifying assumptions will tend to cancel each other out.

EXAMPLE 4.4   Place Limits on Answers

Problem Statement: Estimate the mass of an empty tractor trailer.

Solution: It seems reasonable that the tractor trailer mass should be more than five automobiles, but less than 30 automobiles.

Lower bound = 5 × 3000 lb$_m$ = 15,000 lb$_m$

Upper bound = 30 × 3000 lb$_m$ = 90,000 lb$_m$

Because automobiles typically have a mass between 2000 and 4000 lb$_m$, an average value of 3000 lb$_m$ was used. The estimated mass of the tractor trailer is reasonable; an Internet search revealed that heavy trucks weigh about 42 tons (84,000 lb$_m$).

EXAMPLE 4.5   Extrapolate from Samples

Problem Statement: How much fuel is burned by Texas A&M students for all the Thanksgiving visits home?

Solution: A survey of 20 random students reveals that 6 come from Houston (100 miles away), 4 from Dallas (200 miles), 3 from Austin (90 miles), 2 from San Antonio (190 miles), and 5 live too far to return home for Thanksgiving. Assume that the average car has 1.5 occupants (half have two occupants and half have one occupant). The student population is 42,000 students.

$$\text{Fuel used} = \frac{\text{Total miles}}{\text{Average fuel economy}}$$

$$\text{Total miles} = \text{Houston} + \text{Dallas} + \text{Austin} + \text{San Antonio}$$

$$\text{Houston} = 2 \times \frac{1}{1.5} \times \frac{6}{20} \times 42{,}000 \times 100 \text{ mi} = 1{,}680{,}000 \text{ mi}$$

$$\text{Dallas} = 2 \times \frac{1}{1.5} \times \frac{4}{20} \times 42{,}000 \times 200 \text{ mi} = 2{,}240{,}000 \text{ mi}$$

$$\text{Austin} = 2 \times \frac{1}{1.5} \times \frac{3}{20} \times 42{,}000 \times 90 \text{ mi} = 756{,}000 \text{ mi}$$

$$\text{San Antonio} = 2 \times \frac{1}{1.5} \times \frac{2}{20} \times 42{,}000 \times 190 \text{ mi} = 1{,}064{,}000 \text{ mi}$$

$$\text{Total miles} = 1{,}680{,}000 \text{ mi} + 2{,}240{,}000 \text{ mi} + 756{,}000 \text{ mi} + 1{,}064{,}000 \text{ mi}$$

$$= 5{,}740{,}000 \text{ mi}$$

$$\text{Fuel used} = \frac{5{,}740{,}000 \text{ mi}}{25 \text{ mi/gal}} = 229{,}600 \text{ gal} \approx 230{,}000 \text{ gal}$$

4.6 CREATIVE PROBLEM SOLVING

The Accreditation Board for Engineering and Technology (ABET), in its *1985 Annual Report* (New York, ABET, 1986), defines engineering as "the profession in which a knowledge of the mathematical and natural sciences gained by study, experience, and practice is applied with judgment to develop ways to utilize, economically, the materials and forces of nature for the benefit of mankind." This rather painfully contrived definition (obviously the work of a committee) does, in fact, contain the necessary ingredients to define engineering: skill in mathematics and the natural sciences coupled with a knack for producing useful works with this knowledge. What is missing from the ABET definition is the necessity for creativity and problem-solving skills.

The scientist studies what nature has already created; the engineer creates from nature what did not exist before. In this creative process, the engineer marshals her skill in mathematics, her knowledge of materials, and the particular principles of her specialized engineering discipline, and from this resource creates a new solution for a human need or problem. In almost all cases, the solution is also constrained by economic reality: not only must it answer the particular need that called for a solution, but it must answer the need cheaply.

Somehow, engineering is often stereotyped as dull and stifling work. We are always astounded at how an occupation that builds machines to deliver people to the moon, that has provided means to solve most of the world's problems in transportation, communication, and basic life support, could be cast as dull.

Because creativity figures so centrally in good engineering, it is worthwhile to take another look at the nature of creativity, surely the most misunderstood process of the human intellect. Though possessed by all, many assume that "real" creativity is bestowed upon the few, principally writers and artists. Furthermore, creative expression is perceived to occur in a blinding flash, the creator bringing to life his vision by merely following that which he saw in the creative flash.

Neither of these ideas about creativity is true. Certainly it is true that a new insight will occur in an instant, and a whole new way of looking at things will present itself to the thinker; but this is not the first step or the last in the creative process and often not even the most important step. The greatest artist must spend years perfecting his mastery of the materials used in his field before his flashes of insight can be expressed as works of art. Then the expression of the original thought will often lead to other insights or, at a minimum, numerous creative choices on how to best express the original insight within the framework of skills and materials at hand. Likewise, in engineering, the beginner must master the basics and "practice" engineering before his insights can be brought to life.

This idea brings us back to the second misconception about creativity: that it is the domain of a few artists chosen by a higher power or good fortune. Nothing could be further from the truth. There are two types of problems in the world: those that must be solved all at once and those that can be broken into smaller parts and solved a piece at a time. People who solve the first type of problems are world-class geniuses, and each generation produces one or two such people. Fortunately, most problems are of the second class and lie within the range of the rest of us.

Also, creativity is not a process in which the solution to the problem leaps into existence, but a process in which a myriad of small parts of the whole are solved, until the end product stands complete. This skill at managing small pieces of a larger problem is called by many names; for our discussion, we use the term **problem solving.** Problem-solving skills arise from innate talent, from training, and, most importantly, from long practice. On the whole, persistence and practice are far greater gifts than superior intellect. The world is full of intellectually brilliant failures. On the other hand, few fail who combine perseverance with a bit of good judgment.

This section presents discussions of creativity, followed by a selection of general problems to allow development and practice of problem-solving skills (because we believe that these skills can be taught) with the hope that these general methods will be applied in your later efforts in engineering. Realize that the words *problem solving* have been applied to a long list of activities ranging from simple arithmetic problems to complex open-ended engineering applications. For now we are using the term to mean an activity or thought process that brings to a satisfying conclusion the problems presented.

4.6.1 Creative Leaps, Aha! Experiences, and Sideways Thinking

One of our favorite puzzles was first constructed by Sam Lloyd (1841–1911), who was probably the most prolific puzzlist in America. Sam presented this puzzle to a client who wanted to use it for advertising, and Sam offered to give it to the advertiser for free if he could solve it overnight. The next day Sam received $1000 and showed the advertiser the solution. Here's the puzzle; see if you can save yourself $1000:

Connect the following nine dots with four contiguous straight lines. Draw the four straight lines without lifting your pencil.

Our chief reason for admiring the puzzle is that its solution is without tricks and completely simple and straightforward; but in trying to work it out we block ourselves from seeing the answer. (We've even had a "proof" brought to us that it cannot be done.) During lecture we often use this puzzle to illustrate how we draw rings around ourselves and prevent bringing our full faculties to focus on the problem. With these hints, see if you can do it.

If you solved the puzzle, perhaps you sensed the thrill of accomplishment that we feel when we finally have the necessary insight to see a solution to a problem we've wrestled with. The world we live in has so many first-class problems to tackle, and, for many engineers, a large part of the motivation in life arises from the feelings we experience when solving a particularly knotty problem. If you enjoy solving puzzles, you'll enjoy engineering.

Many books have been written on creative thought, ranging in quality from psychobabble to serious discussions of the nature of creativity. In this text, we can only expose you to a few thoughts on creative endeavors. We will concentrate on an area of creativity called by many names: *Aha! thinking, lateral* or *sideways thinking,* or *flash thinking.* Basically all of these names describe a type of thinking that has little or nothing to do with the linear, logical processes of reasoning. With this type of thinking, the perception of the world is suddenly given a half-twist, and the thinker experiences an insight that allows a solution to the problem at hand.

Following is a series of problems for you to solve, both to give you practice in creative efforts and to analyze and improve the methods you use to creatively solve problems. These first problems are of a class we call **manipulative models.** These problems are similar to working a jigsaw puzzle: basically, you need some physical entity to manipulate in order to solve the problem. Crick and Watson, when they were working out the complicated structure of DNA, had a Tinkertoy model on which to test out their various theories (the model is now in the British Museum of Science). When complicated layouts are to be built (for the control panel for a nuclear reactor or the flight deck of the space shuttle), mockups are made first so that problems can be solved on the model.

So here are your problems:

PROBLEM 4.1

Using six short cylinders (nickels or quarters will do as models) make each touch:

Two and only two others (two completely different ways)
Three and only three others (two completely different ways)
Four and only four others (two completely different ways)

"Touch" is contact between two cylinders either along the curved perimeter or along the flat faces. Here is an example using three nickels and having each touch two and only two others:

1 touches 2 and 3;
2 touches 1 and 3;
3 touches 1 and 2.

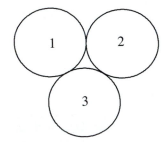

This is also the format for answering the problems: provide a sketch of the solution and a list of each object the other touches. "Completely different ways" means a new physical arrangement that cannot be reached by merely renumbering or rotating an old solution.

PROBLEM 4.2

Continue with a similar exercise, but now use six rectangular **parallelepipeds** (match-boxes or books are appropriate models) and have each touch:

Two others (two completely different ways)
Three others (two completely different ways)
Four others
Five others

The "touch" definition differs for the parallelepipeds. To be touching, a flat surface must be in contact with another flat surface. Contact between corners or between a corner and a flat surface does not count as touching. For example:

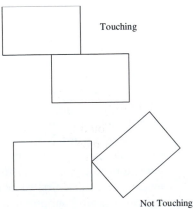

PROBLEM 4.3

Finally, using six equilateral triangles with finite thickness (plane or paper-thin triangles are not the correct shape), make each touch:

Two others (two completely different ways)
Three others (two completely different ways)

As with rectangular parallelepipeds, corner touches do not count.

4.6.2 Classifying Problem-Solving Strategies

A good engineer's time is always spoken for. Because good engineers enjoy engineering, it is indeed an act of love of the profession to take time from their work to write books and articles for the general public. It is even rarer for engineers to analyze how they approach engineering problems and write this down for the student. In 1945, George Polya, a mathematician, published the book *How to Solve It* (Polya, 1945), describing in general terms methods to solve mathematical problems. These methods are well suited to engineering problems. (The Princeton Science Library reissued Polya's book in 2004. It is well worth reading.)

Polya summarized his method as the "How to Solve It List":

First. You have to *understand* the problem.

> *What is the unknown? What are the data? What is the condition?*
>
> Is it possible to satisfy the condition? Is the condition sufficient to determine the unknown? Or is it insufficient? Or redundant? Or contradictory?
>
> Draw a figure. Introduce suitable notation.
>
> Separate the various parts of the condition. Can you write them down?

Second. Find the connection between the data and the unknown. You may be obliged to consider auxiliary problems if an immediate connection cannot be found. You should eventually obtain a *plan* of the solution.

> Have you seen it before? Or have you seen the same problem in a slightly different form?
>
> *Do you know a related problem?* Do you know a theorem that could be useful?
>
> *Look at the unknown!* And try to think of a familiar problem having the same or a similar unknown.
>
> *Here is a problem related to yours and solved before. Could you use it?* Could you use its results? Could you use its method? Should you introduce some auxiliary element in order to make its use possible? Could you restate the problem? Could you restate it still differently? Go back to definitions.
>
> If you cannot solve the proposed problem, try to solve first some related problem. Could you imagine a more accessible related problem? A more general problem? A more special problem? An analogous problem? Could you solve a part of the problem? Keep only a part of the condition, and drop the other part; how far is the unknown then determined, and how can it vary? Could you derive something useful from the data? Could you think of other data appropriate to determine the

TABLE 4.3
Problem-solving strategies

Polya	Woods et al.	Bransford and Stein	Schoenfeld	Krulik and Rudnick
Understand the problem	Define the problem	Identify the problem	Analyze the problem	Read the problem
	Think about it	Define and represent it	Explore it	Explore it
Devise a plan	Plan	Explore possible strategies	Plan	Select a strategy
Carry out the plan	Carry out the plan	Act on the strategies	Implement	Solve
Look back	Look back	Look back and evaluate the effects of your activities	Verify	Look back

unknown? Could you change the unknown or the data, or both if necessary, so that the new unknown and the new data are nearer to each other?

Did you use all the data? Did you use the whole condition? Have you taken into account all essential notions involved in the problem?

Third. *Carry out* your plan.

Carrying out your plan of the solution, *check each step.* Can you see clearly that the step is correct? Can you prove that it is correct?

Fourth. *Examine* the solution obtained.

Can you *check the result?* Can you check the argument? Can you derive the result differently? Can you see it at a glance? Can you use the result, or the method, for some other problem?

There are many variations of this basic scheme; Table 4.3 summarizes strategies suggested by other authors whose work you will find in the bibliography. Most notable is that they all offer roughly the same advice, and that all of them emphasize *checking your results* as the last step. To err is human, but most human of all is to screw everything up in the final steps after the fun of solving the problem is gone and deadlines gallop closer. Many an engineer wishes he had taken an extra 5 minutes to check units or to see if the result really made sense.

Many problem-solving methods have been categorized. The following examples are by no means a complete list—but the particular problem one is pondering has a vanishingly small chance of being categorized anyway. The object of these examples is to get you to think about general methods of attack on problems and, let us hope, to ponder your own methods of tackling problems. Moving from abstract discussions of problem solving to concrete examples will help you to grasp each strategy.

EXAMPLE 4.6   Exploit Analogies or Explore Related Problems

Exploiting analogies is one of the most common approaches to solving a problem. In the old days of engineering (30 years ago), machines called analog computers existed. By using resistors to model fluid friction, capacitors to model holdup tanks, and batteries for pumps, the flow of fluid through a piping system could be modeled with electronics. As you practice engineering, your list of solved problems will grow and become your list against which to test new problems, looking for similarities and analogies.

Problem Statement: Freight carriers charge by rate proportional to the length, depth, and height of a package. What are the minimum lengths, depths, and heights of a rectangular parallelepiped to ship a rod of length q, if the diameter of the rod is negligible? (Adapted from Polya, 1945, and others.)

Solution:

What is the unknown? The lengths of the sides of the parallelepiped, say a, b, and c.
What is the given? The length of the rod q, and that the diameter of the rod is negligible.
Draw a figure.

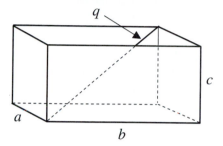

What is the unknown? The diagonal of the parallelepiped.
Do you know any problems with a similar unknown? Yes, sides of right triangles!
Devise a plan.

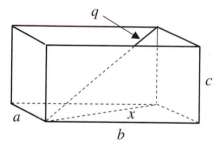

Carry out the plan. If we could find x, then we could relate c and x to q by the **Pythagorean theorem,** that is,

$$q^2 = x^2 + c^2$$

But x^2 is equal to $b^2 + a^2$, so that

$$q = \sqrt{a^2 + b^2 + c^2}$$

Look back. Did we use the appropriate data? Does the answer make sense? (What happens if a goes to zero? Check special cases.) Is there symmetry? (Because a, b, and c can be arbitrarily rotated to any edge, is the form of the answer symmetrical?)

EXAMPLE 4.7   Introduce Auxiliary Elements; Work Auxiliary Problem

Sometimes a problem is too difficult (or it's too early in the morning) to tackle head on. An approach that allows progress and often lights the path to the real insight to unknot the problem is to tackle the problem with one or more of the constraints lifted. By solving the easier problem, we can sometimes maneuver the parameters of the simpler problem until the lifted constraint is also satisfied and we're done. As in this problem, it's spectacular when it works.

Problem Statement: Inscribe a square in any given triangle. Two vertices of the square should be on the base of the triangle, the other two vertices of the square on the other two sides of the triangle, one on each. (From Polya, 1945.)

Solution:

What is the unknown? A method to inscribe a square in any triangle.
What is the given? An arbitrary triangle. Directions for placing square in triangle.

Draw a figure.

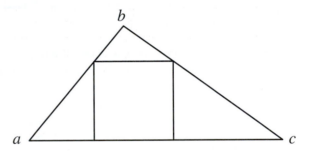

Ouch! This is tough. Even using a graphics program to do the hard part, it still takes us 10 minutes to jockey a square about the right size into a triangle, and then it's only an approximation. With odd-shaped triangles (short height with a long base, or tall with a narrow base), the whole thing gets worse.

Hmmm . . . What if we let squares sit on the base, but they only have to touch the triangle on side *ab*? If we draw a bunch of squares, maybe the lights will come on.

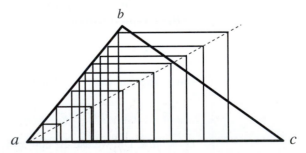

Notice anything unusual about the free vertices? They all lie along a single line! From there, algorithm construction is simple.

Carry out the plan. Construct an arbitrary square on the base of the triangle and draw a line from point *a* on the triangle through the free vertex of the square. Where this line intersects side *bc* on the triangle is where the free vertex of the desired square will coincide with side *bc* of the triangle. Construct a perpendicular line to the base of the triangle through this point. This is one side of the square. Mark off the same distance along the base of the triangle; this is the second side of the square. Construct a perpendicular from this point on the base to side *ab* on the triangle. This locates all four vertices of the square.

Look back. With the clarity of hindsight, why do the vertices lie on a single line? Is this right? Knowing this, can an easier algorithm be developed rather than constructing an arbitrary square on the base of the triangle and drawing a line from point *a* on the triangle through the free vertex of the square?

EXAMPLE 4.8   Generalizing: The Inventor's Paradox

Sometimes it is easier to solve a more general version of a given engineering problem and put in the specific parameters at the end, rather than to solve the particular problem straight away. Much work has been done to solve general problems, so the efforts of others can sometimes be brought to your aid by enlarging your problem. In the study of differential equations, for example, the trick is to identify which general class of equations the particular equation you're working with belongs. Once you've done this, the answer is obvious, because general solutions are given for each class of equations.

Problem Statement: A satellite in circular orbit about the earth, 8347.26 miles from the center of the earth, moves outward 6.1 ft. How much new area is enclosed in its new orbit?

Solution:

What is the unknown? The extra area inside the orbit.
What is the given? Circular orbit with a radius of 8347.26 miles. Move out 6.1 ft.
Draw a figure.

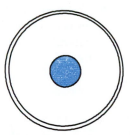

Make a plan. Rather than dive in and crunch numbers, let's do a bit of algebra for a second.
Carry out the plan. Let R be the beginning radius, and let q be the distance added to the radius. For the smaller orbit, the area enclosed was πR^2. For the new orbit, the area enclosed is $\pi(R + q)^2$. The difference is the new area enclosed, or

$$\text{New area} = \pi(R + q)^2 - \pi R^2 = \pi(R^2 + 2Rq + q^2) - \pi R^2$$

$$= \pi(2Rq + q^2) = \pi q(2R + q)$$

$$= 3.14159\left(6.1 \text{ ft} \times \frac{\text{mile}}{5280 \text{ ft}}\right)\left[2(8347.26 \text{ mi}) + \left(6.1 \text{ ft} \times \frac{\text{mile}}{5280 \text{ ft}}\right)\right]$$

$$= 61 \text{ square miles}$$

Look back. Is this reasonable? How could this result be checked? Are the units right? Are the conversion factors right? If q went to zero, does the formula work? If R went to zero?

    If you are not convinced that this algebraic approach is better, start right out with numbers and see how you do. Besides that, by solving this a general way, several avenues open up for checking the result.

EXAMPLE 4.9   Specializing; Specializing to Check Results

Specialization is our all-time favorite tool for problem solving. For checking results as the final step in problem solving, specialization is the tool of choice. Results are often known for special cases, and this offers a wonderful way to check a general algorithm. Another standard approach is to push the algorithm's parameters to their limits and assure that the algorithm fails gracefully or scoots to infinity if it's supposed to.

Problem Statement: In a triangle, let r be the radius of the inscribed circle, R the radius of the circumscribed circle, and H the longest altitude. Then prove

$$r + R \le H$$

(In Polya, 1945, quoting from *The American Mathematical Monthly* 50, 1943, p. 124 and 51, 1944, pp. 234–236.)

Solution:

What is the unknown? How to show that the sum of the radii of inscribed and circumscribed circles is less than or equal to the greatest height of the triangle.
What is the given? The formula and directions for constructing the figure.
Draw a figure.

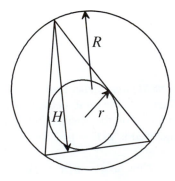

Devise a plan. This looks tough to prove. It's likely that $R + r$ is less than H in this example, but is it always true? Let's go to extremes. (That's where all the fun is anyway, right?) Let's look at a narrow-based, tall triangle and then a broad-based, short triangle. *Carry out the plan.*

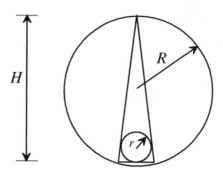

From the figure and with a bit of thought, we see that as the triangle gets taller, r will approach zero, and H will approach $2R$, so indeed H is greater than R. How about the other way?

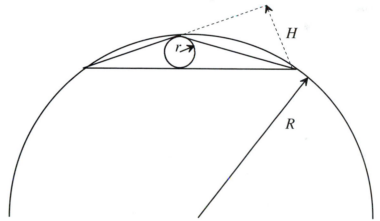

But hang on a minute. Going this way, R is much larger than H, so the conjecture cannot be true.

Look back. Is there another way to prove that broad-based triangles don't work? Couldn't the tall, narrow triangle be turned over to become a short, broad triangle? What happens as the base gets infinitely long? When does $R + r = H$?

EXAMPLE 4.10   *Decomposing and Recombining*

An engineer's most-used dictum is, Divide and conquer. Big problems are broken into little problems and solved one at a time, or parceled out among a group to be recombined for the finished product. This problem (from Schoenfeld, 1985) gives a sample of that approach.

Problem Statement: Given that p, q, r, and s are real positive numbers, prove that

$$\frac{(p^2 + 1)(q^2 + 1)(r^2 + 1)(s^2 + 1)}{pqrs} \geq 16$$

Solution:

What is the unknown? How to show that the formula is always equal to or greater than 16.
What is the given? The formula.
Draw a figure. Don't know how to yet.
Devise a plan. As a first effort, we try numbers at random for p, q, r, and s. If we can find a contradiction, we're done. But, no luck. In fact, usually the numbers are a lot bigger than 16. Hmmm . . . Isn't there symmetry in the formula? Really, isn't $(p^2 + 1)/p$ the same formula as $(q^2 + 1)/q$, $(r^2 + 1)/r$, and $(s^2 + 1)/s$? Then, if I can solve this small piece, I understand the whole formula. Now,
Carry out the plan.
Draw a figure.

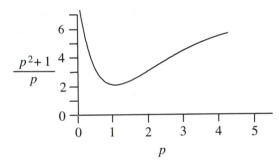

Look back. Now it is clear how the formula works. Each piece will never be smaller than 2, and because $2 \times 2 \times 2 \times 2 = 16$, this is the minimum for the formula as a whole. How could this be checked? What happens at zero, the smallest positive number put into the formula? Can you prove that 2 is a minimum value for the formula?

EXAMPLE 4.11   **Taking the Problem as Solved**

While taking the abstract approach is often useful (see the Inventor's Paradox, Example 4.8), the brain usually works better on concrete objects. If stuck on a problem, try to visualize the answer, no matter how approximate that vision may be. From this vision, try to work backward or deduce what is missing between the starting point and the solution.

Problem Statement: Given two intersecting straight lines and a point P marked on one of these lines, and using a straightedge and a compass, construct a circle that is tangent to both lines and passes through point P as one of its points of tangency. (From Schoenfeld, 1985.)

Solution:

What is the unknown? How to construct the circle.
What is the given? Two intersecting lines and one point of tangency.
Draw a figure.

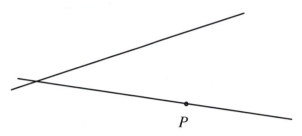

Devise a plan. Let's sketch in something that's close to a circle and close to the right size and see what happens.

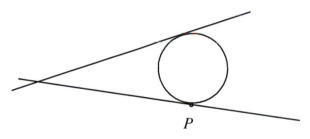

Well, not too bad for a sketch, but now what? The center of the circle looks straight up from point *P*. But it shouldn't be, right? Aren't tangents at right angles to the radius of the circle at the point of tangency? Well, that helps. I can construct a perpendicular at point *P,* but I still can't draw the circle because I don't know where the center is. Hmmm . . .

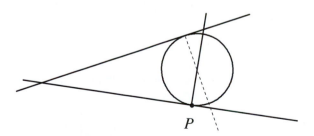

If I only knew where to place the other perpendicular from the other intersecting line (the dotted line in the figure), I'd be done. But I don't. Bummer. . . . But wait a minute! If that line *were* there, then the radius from the center to point *P* is the same distance as the radius to the tangency on the other line, right?

That means that the center of the circle is on the angular bisector, thus

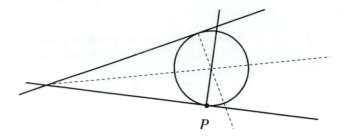

P

So now the scheme is clear. At point *P,* construct a perpendicular. Bisect the angle between the two intersecting lines. The point where the bisector and perpendicular meet is the center of the circle.

Look back. Is this true for any two intersecting lines? How about obtuse angles? Can we do that? How about as the lines intersect at angles near zero (i.e., as the lines approach being parallel)?

EXAMPLE 4.12   Working Forward / Working Backward

One of our students informed us that learning this trick alone was worth all the effort of the entire course. His work is synthesizing organic compounds, and this is his standard trick. He looks at what he wants to make and then mentally breaks the compound down to simpler components that he can make or, better yet, has sitting on his shelf. Another less esoteric application of this method is in solving maze puzzles. Start at the end and work back toward the start. Often the puzzle is easier this way. What usually happens in practice is that one starts from the beginning and works forward until stymied, then starts at the end and works backward. With luck one meets oneself somewhere in the middle.

Problem Statement: Measure exactly 7 ounces of liquid from a large container using only a 5-ounce container and an 8-ounce container. (Practiced with libations at many student hangouts.) (From Polya, 1945, and many others.)

Solution:

What is the unknown? How to measure seven ounces.
What is the given? Unlimited liquid, 5-oz and 8-oz containers.
Draw a figure.

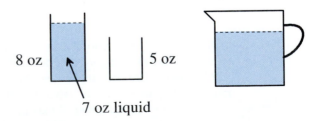

8 oz   5 oz

7 oz liquid

This is, of course, the final state we seek: the 7 oz of liquid in the 8-oz container. (It's hard to put that much liquid into the 5-oz container.)

Devise a plan. The obvious comes to mind first. We can get 3 oz by filling the 8-oz container and then pouring from it into the 5-oz container until the 5-oz is full, leaving 3 oz of liquid in the 8-oz container. Also, we can get 2 oz by filling the 5-oz container, pouring it into the 8-oz container, filling the 5-oz container again, and emptying it into the 8-oz container until the 8-oz container is full, leaving 2 oz in the 5-oz container. Well, so what? Let's start at the end and see if we can meet in the middle.

Carry out the plan. If we back up one step from the final state we would have:

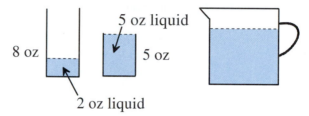

Bingo! I know how to get 2 oz. So now we perform the forward sequence to get 2 oz, then empty the full 8-oz container, pour in the 2 oz, refill the 5-oz container, and empty it into the 8-oz container.

Look back. Is each step correct? Anything special about 8-, 7-, and 5-oz measurements?

EXAMPLE 4.13   Argue by Contradiction; Using *Reductio ad Absurdum*

Occasionally, nature is kind enough to present a problem that must have only one of a small number of answers. It may be impossible to prove directly that one particular answer is correct. However, if you can prove that all others are not the answer, you prove indirectly that the remaining choice is correct.

Problem Statement: Write numbers using each of the 10 digits (0 through 9) exactly once such that the sum of the integers is exactly 100. For example, 29 + 10 + 38 + 7 + 6 + 5 + 4 = 99 uses each digit once, but sums to 99 instead of 100. And then there's 19 + 28 + 31 + 7 + 6 + 5 + 4 = 100, but . . . no zero. (From Polya, 1945.)

Solution:

What is the unknown? How to create numbers using the digits 0 through 9 so that the sum is 100.

What is the given? The digits 0 through 9, examples.

Draw a figure. For once, this doesn't help.

Devise a plan. After a lot of trial and error, one is convinced that the answer is not obvious, if it exists. However, proof that it is impossible is not obvious either. We'll bet it has something to do with which digits are in the tens place. Let's analyze how the sums can be made and, assuming it can be done, look for a contradiction.

Carry out the plan. First, the sum of the 10 digits, each taken singly, is:

$$0 + 1 + 2 + 3 + 4 + 5 + 6 + 7 + 8 + 9 = 45$$

If we add "0" to any of the digits, for example we change 1 to 10, the sum increases by 10 times whichever digit we used, less the digit itself. For 10, the sum is $45 + 10 - 1 = 54$. For 40, the sum is $45 + 40 - 4 = 81$. For 60, the sum is 99; for 70 the sum is 108, too large, as are 80 and 90. If digits are combined with numbers other than 0, for example combining the 4 and the 2 to make 42, the sum is $45 - 4 + 40 = 81$. Ahhh, only the digits used in the tens place are "lost" in the sum. We don't know how many digits will be lost (at most three, right?), but we can write an equation about their sum:

$$10 \times (\text{tens\_sum}) + 45 - \text{tens\_sum} = 100$$

Now we're cooking. But wait—solving the equation for tens_sum, we get 55/9? What kind of junk is this? Exactly the kind of junk we hoped for. Because there is no way to add integers and get 55/9, then the conjecture that the digits can sum to 100 must be false.

Look back. Is there some way to prove our method is correct? How about if we relax the requirement that the digits be integers? Can we use the digits and get 100?

4.7   SUMMARY

During the 40 or so years you will practice engineering, technology will keep changing dramatically. Therefore, it is impossible for your professors to teach you every fact you will need; today's state-of-the-art fact will become obsolete tomorrow. Instead, to prepare you for the future, we can help train your mind to think and solve problems. These skills never become obsolete.

Engineers face a variety of problems, including research, troubleshooting, mathematics, resource, and design problems. The problem-solving process can start once the problem is identified. To solve problems, engineers apply both synthesis (where pieces are combined together into a whole) and analysis (where the whole is dissected into pieces).

During your schooling, you will be confronted with thousands of problems. So you can check your work, the solution to these problems is provided to you. If you make a mistake, the consequences are not severe—the loss of a few points on a homework assignment or exam. In the real world, there are no answers in the back of the book, and the consequences of making a mistake can be catastrophic. You are well-advised to develop a systematic problem-solving strategy that leads you to the correct answer. We offer a suggested approach in the section on "error-free" problem solving.

One of the most valuable skills in engineering is the ability to estimate answers from incomplete information. Because we almost never have all the information we need to precisely solve a problem, nearly every engineering problem can be viewed as an estimation problem; they differ only in the degree of uncertainty in the final answer.

Many authors agree that problem solving can be broken into four or five steps: (1) understand the problem; (2) think about it; (3) devise a plan; (4) execute the plan; and (5) check your work. While thinking about the problem, a number of **heuristic** approaches (i.e., hints) can lead you to the solution. For example, you can exploit analogies, introduce auxiliary elements, generalize, specialize, decompose, take the problem as solved, work forward/backward, or argue by contradiction.

Perhaps you enjoy solving puzzles in the magazines or online. Although these may seem like trivial games, these puzzles can actually train your mind to solve engineering problems. Just as a boxer may train with a jump rope for an upcoming fight, an engineer can train with puzzles to prepare for the main event: solving engineering problems.

Further Readings

Albrecht, K. *Brain Power: Learning to Improve Your Thinking Skills.* New York: Prentice Hall, 1987.

Bransford, J. D., and B. S. Stein. *The Ideal Problem Solver,* 2nd ed. New York: Worth Publishers, 1993.

de Bono, E. *The Five-Day Course in Thinking.* New York: Pelican, 1972.

———. *New Think.* New York: Avon, 1985.

———. *Lateral Thinking: Creativity Step by Step.* New York: Harper, 1973.

Epstein, L. C. *Thinking Physics,* 3rd ed. San Francisco: Insight Press, 2002.

Gardner, M. *New Mathematical Diversions.* New York: Mathematical Association of America, 1995.

———. *Mathematical Carnival.* New York: Alfred A. Knopf, 1975.

———. *Mathematical Circus.* New York: Mathematical Association of America, 1996.

Graham, L. A. *The Surprise Attack in Mathematical Problems.* New York: Dover Publications, 1968.

Krulik, S., and J. A. Rudnick. *Problem Solving: A Handbook for Senior High School Teachers.* Boston: Allyn and Bacon, 1988.

Miller, J. S. *Millergrams 1: Some Enchanting Questions for Enquiring Minds.* Sydney: Ure Smith, 1966.

Moore, L. P. *You're Smarter Than You Think.* New York: Holt, Rinehart & Winston, 1985.

Polya, G. *How to Solve It.* Princeton, NJ: Princeton University Press, 2004.

Row, T. S. *Geometric Exercises in Paper Folding.* Ann Arbor, MI: University of Michigan, 2005.

Sawyer, W. W. *Mathematician's Delight.* New York: Penguin USA, 1991.

Schoenfeld, A. H. *Mathematical Thinking and Problem Solving.* Mahwah, NJ: Erlbaum, 1994.

Schuh, F. *The Master Book of Mathematical Recreations.* New York: Penguin, 1991.

Stewart, I. *Game, Set, and Math: Enigmas and Conundrums.* Oxford: Basil Blackwell, 1989.

Witt, S. *How to Be Twice as Smart: Boosting Your Brainpower and Unleashing the Miracles of Your Mind.* New York: Penguin, 2002.

Woods, D. R., J. D. Wright, T. W. Hoffman, R. K. Swartman, and I. D. Doig. "Teaching Problem Solving Skills." *Engineering Education,* December 1975, pp. 238–243.

PROBLEMS

4.1 Mary Mermaid is taking swimming lessons in a circular pool. Mary starts at one edge of the pool and swims in a straight line for 12 meters, where she hits the edge of the pool. She turns and swims another 5.0 meters and again hits the edge of the pool. As she is examining her various scrapes, she realizes that she is exactly on the opposite side of the pool from where she started. What is the diameter of the pool?

4.2 What is the shortest path for an ant crawling on the surface of a unit cube from the starting point to the ending point shown:

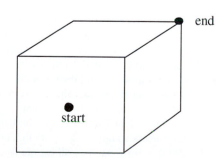

4.3 What is the shortest path for an ant crawling on the surface of a unit cube from the starting point to the ending point shown:

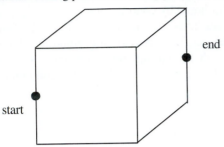

4.4 A string is fit snugly around the circumference of a spherical hot air balloon. More hot air is added (probably by a prominent scientist's lecture) and it now takes an additional 12.4 feet of string to fit around the circumference. What is the increase in diameter?

4.5 The poet Henry Wadsworth Longfellow, in his novel *Kavenaugh,* presented the following puzzle:

When a water lily stem is vertical, the blossom is 10 cm above the water. If the blossom is pulled to the right keeping the stem straight, the blossom touches the water 21 cm from where the stem came through the water when vertical. How deep is the water? (*Hint:* See figure.)

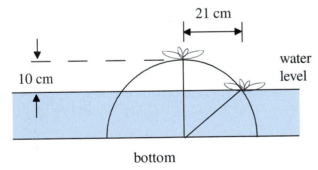

4.6 A farmer goes to market with $100 to spend. Cows cost $10 each, pigs cost $3 each, and sheep cost half a dollar each. The farmer buys from the cattle dealer, the pig dealer, and the sheep dealer. She spends exactly $100 and buys exactly 100 animals. How many of each animal does she buy?

4.7 The king and his two children are imprisoned at the top of a tall tower. Stone masons have been working on the tower and have left a pulley fixed at the top. Over the pulley runs a rope with a basket attached to either end. In the basket on the ground is a stone like the ones used to build the tower. The stone weighs 35 kg$_f$ (75 lb$_f$). The king figures out that the stone can be used as a counterbalance—provided that the weight in either basket does not differ by more than 7 kg$_f$ (15 lb$_f$). The king weighs 91 kg$_f$ (195 lb$_f$), the princess weighs 49 kg$_f$ (105 lb$_f$), and the prince weighs 42 kg$_f$ (90 lb$_f$). How can they all escape from the tower?

(They can throw the stone from the tower to the ground!) (Attributed to Lewis Carroll.)

4.8 From Problem 4.7, add a pig that weighs 28 kg$_f$ (60 lb$_f$), a dog that weighs 21 kg$_f$ (45 lb$_f$), and a cat that weighs 14 kg$_f$ (30 lb$_f$). There is an extra limitation: there must be one human at the top and bottom of the tower to put the animals in and out of the basket. How can all six escape?

4.9 Calculate the ratio of the area to the volume for a unit cube, a unit sphere inscribed inside the cube, and a right cylinder inscribed inside the cube.

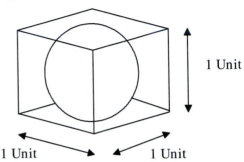

Next, for each having a unit volume (i.e., all three solids have the same volume) calculate the area-to-volume ratios for a sphere, cube, and cylinder.

4.10 Dr. Bogus, a close friend of ours, during a long session at the Dixie Chicken restaurant, was doodling on a napkin and "proved" that all numbers are equal. This came as quite a surprise to us, and we have delayed his calling the president only until you can review his proof. Here is a translation of the napkin doodles:

Pick two different numbers, a and b, and a nonzero number c that is the difference between a and b, thus:

$$a = b + c \qquad c \neq 0 \qquad (1)$$

Multiply both sides by $a - b$

$$a(a - b) = (b + c)(a - b) \qquad (2)$$

or

$$a^2 - ab = ab - b^2 + ac - bc \qquad (3)$$

Subtract ac from both sides,

$$a^2 - ab - ac = ab - b^2 - bc \qquad (4)$$

Factor out a from the left side of the equation, and b from the right,

$$a(a - b - c) = b(a - b - c) \qquad (5)$$

Eliminate the common factor from both sides, and

$$a = b \qquad (6)$$

Because a and b can be any number, c can be positive or negative; thus, all numbers are equal to each other. Exactly which step(s) are wrong in the proof above, and why?

4.11 Dr. Bogus, now on a roll, also presented us with a proof that all our attempts to measure area are wrong. Again, we were somewhat startled, but he showed us that by simply rearranging the space within an area, one could get different answers. He presented us with the following drawing:

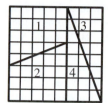

A square 8 units on a side is cut into 4 pieces. The pieces are then rearranged into a 5×13 rectangle.

Wait a minute!

$8 \times 8 = 64$

$5 \times$

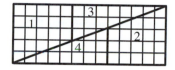

Where does the extra square come from? Help us, or Dr. Bogus will get a Nobel Prize before we do.

4.12 What is the plane angle between Line *A* and Line *B*, drawn on a unit cube?

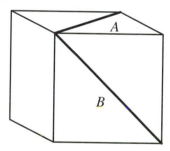

4.13 A fly is at the midpoint of the front edge of a unit cube as shown in the figure. What minimum distance must it crawl to arrive at the midpoint of the opposite top edge?

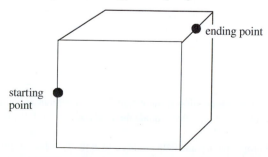

4.14 Suppose you wish to average 40 mph on a trip and find that when you are half the distance to your destination you have aver-

aged 30 mph. How fast should you travel in the remaining half of the trip to attain an overall average of 40 mph?

4.15 A ship leaves port at 12:00 P.M. (noon) and sails east at 10 miles per hour. Another leaves the same port at 1:00 P.M. and sails north at 20 miles per hour. At what time are the ships 50 miles apart?

4.16 At a certain Point *A* on level ground, the angle of elevation to the top of a tower is observed to be 33°. At another Point *B*, in line with *A* and the base of the tower and 50 feet closer to the tower, the angle of elevation to the top is observed to be 68°. Find the height of the tower.

4.17 Using a straight track with 10 spots, set four pennies in the leftmost spots and four dimes in the rightmost spots, leaving the center two spots blank, as shown in the figure:

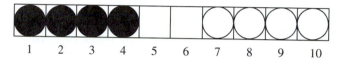

The object of the game is to move the pennies to the rightmost spots and the dimes to the left. You can move coins only by jumping a single coin into a vacant spot or by moving forward one to a vacant spot. No backward moves are allowed: all pennies must move to the right; all dimes to the left. Number the squares as shown to describe your solution.

4.18 Jack Jokely got up late one morning and was trying to find a clean pair of socks. He has a total of six pairs (three maroon and three white) in his drawer. Jack is not a very good housekeeper and his socks were just thrown in the drawer and, to make the morning perfect, the lightbulb is burned out so he can't see. How many socks does he have to pull out of the drawer before he gets a matched pair (two maroon socks or two white socks)?

4.19 A farmer has a piece of land he wants to give to his four sons (see figure below). The land must be divided into four equal-sized and equal-shaped pieces. How can the farmer accomplish his task?

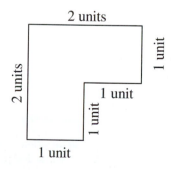

4.20 A peasant had the wonderful good fortune to save the life of the King of Siam. The King, in turn, granted her payment in any form she wished. "I am a simple woman," said the peasant; "I wish merely to never be hungry again." The peasant asked to have payment made onto a huge chessboard painted on the floor. Payment was to be as follows: one grain of rice on the first square, two grains on the second square, four grains on the third square, eight onto the fourth, etc., until all 64 squares were filled. How many grains of rice will the peasant receive from the King? Roughly what volume of rice is this?

4.21 In a field there are cows, birds, and spiders. Spiders have four eyes and eight legs each. In the field there are 20 eyes and 30 legs. All three animals are present, and there is an odd number of each animal. How many spiders, cows, and birds are present? (From Gardner, 1978.)

4.22 Florence Fleetfoot and Larry Lethargy race on a windy day: 100 yards with the wind, they instantly turn around, and race 100 yards back against the wind. Larry is unaffected by the wind, but Florence goes only 90% as fast running against the wind as when running with no wind. Running with the wind improves Florence's no-wind speed by 10%. On a day with no wind, Florence and Larry tie in a 100-yard race. Who wins the windy-day race, and by how much?

4.23 Given a fixed triangle T with base B, show that it is always possible to construct with a straightedge and compass, a straight line parallel to B dividing the triangle into two parts of equal area. (From Schoenfeld, 1985.)

4.24 A pipe one mile in length and with 1-inch inside diameter is set in the ground such that it is optically straight, that is, a laser beam will pass down the centerline of the whole pipe. The centerline of the pipe is exactly level, that is, it is at right angles with a line passing through the center of the earth. Now, by means of a funnel and rubber tubing, water is poured into one end of the pipe until it flows from the other end. The funnel and tubing are removed and water is allowed to flow freely from the open ends. Neglect the surface tension of the water (it has little effect on this problem anyway) and calculate how much water is left in the pipe (say to within 10%). (*Hint:* The answer is not zero, and take the radius of the earth to be exactly 4000 miles should you need that information. *Further hint:* This problem is tougher than the rest of 'em put together.)

4.25 Estimate the number of toothpicks that can be made from a log measuring 3 ft in diameter and 20 ft long.

4.26 Estimate the number of drops in the ocean. Compare your estimate to Avogadro's number.

4.27 Estimate the maximum number of cars per hour that can travel down two lanes of a highway as a function of speed (i.e., at 50, 60, and 70 mph). For safety reasons, the cars must be spaced one car length apart for each 10 mph they are traveling. For example, if traffic is moving at 50 mph, then there must be five car lengths between each car. A city councilman proposes to solve traffic congestion by increasing the speed limit. How would you respond to this proposal?

4.28 If electricity costs $0.07/kWh (kilowatt-hour), how much money does a typical household spend for electricity each year?

4.29 Estimate the number of books in your university's main library.

4.30 Estimate the amount of garbage produced by the United States each year.

4.31 Estimate the amount of gasoline consumed by automobiles in the United States each year. If this gasoline were stored in a single tank measuring 0.1 mi by 0.1 mi at the base, what is its height?

4.32 A vertical 10-ft column must support a 1,000,000-lb_f load located at the top of the column. An engineer must decide whether to construct it from concrete or steel. She will select the column that is lightest. Assume the column will be designed with a safety factor of 3, meaning the constructed column could support a load three times heavier, but no more. Estimate the mass of a steel column and a concrete column.

4.33 Estimate the number of party balloons it would take to fill your university's football stadium to the top.

4.34 Estimate the amount of money students at your university spend on fast food each semester.

4.35 Estimate the mass of the air on planet earth. What fraction of the total earth mass is air?

4.36 Estimate the mass of water on planet earth. What fraction of the total earth mass is water?

4.37 Estimate the length of time it would take for a passenger jet flying at Mach 0.8 ($\frac{8}{10}$ the speed of sound) to fly around the world. Make allowances for refueling.

4.38 Using Archimedes' principle, estimate the mass that can be lifted by a balloon measuring 30 ft in diameter. The temperature of the air in the balloon is 70°C, and its pressure is 1 atm.

Glossary

application A process whereby appropriate information is identified for the problem at hand.
Archimedes' principle The total mass of a floating object equals the mass of the fluid displaced by the object.
comprehension The step in which the proper theory and data are used to actually solve the problem.
conversion factor A numerical factor that, through multiplication or division, converts a quantity expressed in one system of units to another system of units.
engineering analysis paper A paper with a light grid pattern that can be used for graphing or drafting.

heuristic The use of speculative strategies for solving a problem.

iterative procedure Repetition of a sequence of steps to solve a problem.

manipulative models A class of problems that have some physical entity that can be used to one's advantage in order to solve a problem.

parallelepiped A solid with six parallelogram faces that are parallel to the opposite face.

problem solving A process in which an individual or a team applies knowledge, skills, and understanding to achieve a desired outcome in an unfamiliar situation.

Pythagorean theorem The sum of the squares of the lengths of the sides of a right triangle is equal to the square of the length of the hypotenuse.

reductionism The ability to logically break down a problem into pieces.

CHAPTER 5

Introduction to Design

The ability to create something out of nothing makes design one of the most exciting aspects of engineering. To be successful, design engineers require a broad set of talents, including knowledge, creativity, people skills, and planning ability. As we briefly described in Chapter 2, design engineers follow the *engineering design method.*

In this chapter, we examine the engineering design method in detail. Although there are many variations, we will use the steps illustrated in Figure 5.1. The first four steps of the engineering design method proceed straightforwardly. The next four steps are repeated three times: The first pass is a **feasibility study** where ideas are roughed out, the second pass is a **preliminary design** where some of the more promising ideas are explored in more detail, and the third pass is a **detailed design** where highly detailed drawings and specifications are prepared for the best design option. Finally, the last two steps proceed straightforwardly. The end result of the engineering design method is a product, service, or process that serves the needs of humankind. Often, after the method is complete, the end result can still be improved, so the process is repeated by returning to the first step.

The engineering design method contains the following elements:

- *Synthesis*—combining various elements into an integrated whole.
- *Analysis*—using mathematics, science, engineering techniques, and economics to quantify the performance of various options.
- *Communication*—writing and oral presentations.
- *Implementation*—executing the plan.

The engineering design method is often an *iterative* procedure, meaning some of the steps must be repeated because information needed at the beginning is not known until later steps are completed. Further, the engineering design method is not a rigid procedure to be slavishly followed; rather, it is a general guide.

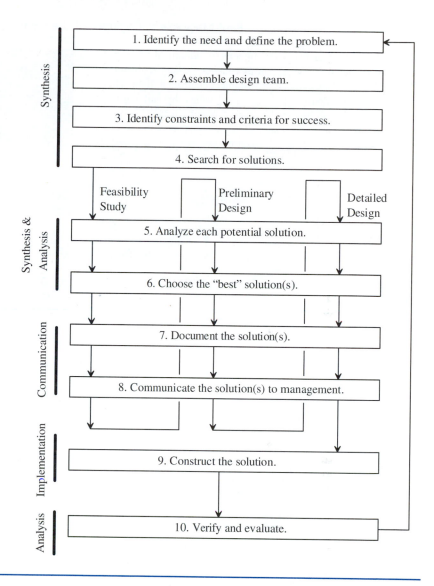

FIGURE 5.1
The engineering design method.

5.1   THE ENGINEERING DESIGN METHOD

Let us examine each step in the engineering design method in detail.

5.1.1 Step 1: Identify the Need and Define the Problem

An engineer is a person who applies science, mathematics, and economics to meet the needs of humankind. Therefore, the job of an engineer starts when a need is identified.

Needs may be identified by a creative engineer who goes through life saying, "There must be a better way." The military may identify a need when its intelligence sources reveal that the enemy has a new capability, and a countermeasure must be developed. (Of

Paul MacCready, Engineer of the Century

In 1977, Paul MacCready became famous for winning the $95,000 Kremer Prize for the first person to construct a heavier-than-air, human-powered plane capable of sustained, controlled flight. This prize stood unawarded for 18 years. Paul won it by building the *Gossamer Condor,* an innovative plane made of advanced materials and employing a sophisticated aerodynamic design.

Pterodactyl in flight.

Gossamer Condor.

British industrialist Henry Kremer then upped the stakes and offered a $213,000 prize to the first person to construct a heavier-than-air, human-powered plane capable of crossing the English Channel. In 1979, MacCready's *Gossamer Albatross* successfully met this challenge.

In 1981, MacCready constructed the *Solar Challenger,* which carried a pilot 163 miles at 11,000 feet in a craft powered solely by sunlight.

GM Sunny Racer.

Solar Challenger.

In 1984, his human-powered *Bionic Bat* won two Kremer Prizes for speed. Later, he developed a radio-controlled, wing-flapping replica of a giant pterodactyl, which was featured in the Imax film *On the Wing.*

In 1987, he built the *Sunny Racer,* a solar-powered car that won an Australian race by going 50% faster than his nearest competitor.

In 1990, he introduced the *Impact* electric automobile, which was mass-produced by General Motors as the EV1. To break the stodgy image of electric cars, this automobile was designed with performance in mind.

Paul MacCready founded AeroVironment, Inc., a company involved with air quality, hazardous wastes, alternative energy, and efficient vehicles for land, sea, and air.

In recognition of his outstanding achievements, Dr. MacCready has received many awards including the *Engineer of the Century Gold Medal* offered by the American Society of Mechanical Engineers.

Impact electric car, prototype of the General Motors EV1.

Source: Information provided by AeroVironment, Inc.

Photos courtesy Paul MacCready/Mark Holtzapple.

course, the enemy will soon learn of our countermeasure, so they will develop a counter-countermeasure. We will counter with a counter-counter-countermeasure, and so on.) Some needs are identified by management or sales personnel who are very familiar with the market and can spot a need for a new product. Government regulations may create needs by setting new standards for safety or pollution control. Politicians may create needs by promising their constituents new roads or buildings. The growing world population creates stresses on our environment and creates new needs to reduce the stress.

Notice that often the engineer does not identify the need; rather, she is there to serve humankind when the need has been identified by others.

Once the need is identified, the problem must be defined. Without this step, we might solve the wrong problem. For example, suppose that a highway is congested and causing troublesome delays for commuters. Once identified, we might *define* the problem as, "How do we widen the road to accommodate more traffic?" However, experience has shown that widening the road often results in larger traffic jams, because commuters soon learn of the extra capacity and swarm to the wider road. Perhaps the problem is not with the road. Rather, perhaps there is a need for a commuter railway, or a high-occupancy lane that allows cars with multiple passengers to travel more quickly. Perhaps a better statement of the problem is, "How do we create a transportation system to move more people quickly and efficiently?"

5.1.2 Step 2: Assemble Design Team

Because of the complexity of modern engineering projects, rarely is a design tackled by a single individual; rather, design is done with teams of individuals who have complementary skills. In the past, a design problem was tackled by specialists who worked in a compartmentalized manner via **sequential engineering**. For example, suppose an automobile was being designed. First, the stylists would decide on the body shape, then the mechanical engineers would determine how to form the body panels and fit the engine under the hood. Then the electrical engineers would design the electrical system, then the production engineers would design the production line, and finally the marketers would develop an advertising campaign. Although sequential engineering worked, it was not optimal. Each specialist could find a **local optimum** for each step, but only within the constraints of what they were given by the previous specialist. Such an approach missed the **global optimum.**

To find the global optimum requires the specialists to work together right from the beginning using an approach called **concurrent engineering.** To illustrate the benefits of concurrent engineering, suppose that while the automobile is being conceptualized, the marketer and the stylist worked together to establish a highly salable design. Further, the mechanical engineer would be involved so that the body style can accommodate the engine. Also, suppose the design objectives can be achieved only by using new materials, such as aluminum space frames or polymer body panels. Obviously, the production engineer must be involved because these new materials will make a big impact on the manufacturing methods. Further, suppose the design objectives can be met only by using a hybrid engine, in which a gasoline engine provides baseline power but an electric motor provides peak power during rapid accelerations. Clearly, with this design, the electrical system is an integral part of the automobile and cannot be designed as an afterthought.

5.1.3 Step 3: Identify Constraints and Criteria for Success

Every project has constraints or limitations, because resources are never infinite. The constraints must be identified early, as they affect the project planning. Typical constraints are listed here:

- *Budget.* Before a project is initiated, the engineers must know the proposed budget, because it affects the resources they can amass for the project.
- *Time.* Some projects must be completed quickly, because the need is urgent. The engineers must know the time allocated to the project, because it determines the number and type of options they can consider.
- *Personnel.* As the project team is assembled, the engineer must know the number of people assigned to the team and their skills. A large budget with ample time does not forecast success unless skilled individuals are working on the project.
- *Legal.* Legal constraints can be restrictive in today's world. Before a large project is undertaken, it must be coordinated with numerous government bureaucracies, each with a different responsibility (water pollution, air pollution, sewage, traffic control, etc.). Legal constraints can cause severe delays and cost overruns if not entered into the planning process.
- *Material properties and availability.* Engineers are always constrained by material properties. For example, it is well known that engine efficiency improves as higher temperatures are employed. However, we are constrained by materials capable of withstanding the high temperatures. Perhaps laboratories have developed new materials (e.g., ceramics) with the needed properties, but until they are available on a commercial scale, they are of no use to the project.
- *Off-the-shelf construction.* Early in the project, the engineers must understand whether they are restricted to assembling off-the-shelf components, or whether they are permitted to custom-design the required equipment. Off-the-shelf items are available quickly and are well tested; however, they may compromise ultimate success if they are not customized to the project requirements.
- *Competition.* The engineers must understand whether the ultimate product is a unique item or whether it will compete against other similar products.
- *Manufacturability.* Many items can be made in small quantities in a laboratory or machine shop, but may be unsuitable for mass production. For example, fighter jets can use high-performance, exotic, lightweight materials (e.g., graphite fiber composites) because they are hand-built in small numbers (maybe 50 per year). These materials are not suitable for automobiles, which are produced in higher quantities (maybe 100,000 of a given model per year).

Once the constraints of the project have been identified, it is necessary to determine the criteria for success; that is, what are the goals for the project? Engineering projects have many goals, some of which are listed here:

- *Aesthetics.* With consumer products, aesthetics play a large role in success. It does not matter how rugged, reliable, or functional the product is; if it is ugly, it will not sell. Aesthetics are difficult to define and are highly subjective, but a product that is well balanced and in proportion and has coordinated colors will usually have aesthetic appeal. An important aesthetic principle is that form follows function, meaning that each component of the product serves a useful purpose. Products that violate this principle are often short-lived. For example, automobiles in the 1950s and 1960s sometimes had large fins added for styling purposes. Because these fins had no function, they were a passing fad that thankfully has not reappeared.

- *Performance.* The performance of a product is generally determined by the producer, unless the project responds to a specific customer request. For example, the producer may specify the following automobile performance: 0 to 60 mph acceleration = 10 s, 60 to 0 mph braking = 150 ft, and fuel economy = 25 miles per gallon.

- *Quality.* The quality of a product is determined by the consumer. Quality is often defined as "fitness for use." For example, a consumer may expect an automobile to have the following qualities: 0 to 60 mph acceleration = 6 s, 60 to 0 mph braking = 110 ft, and fuel economy = 40 miles per gallon. An automobile that does not meet these criteria will be deemed of low quality even if it is reliable and stylish.

- *Human factors.* Because most products are used by humans, successful products must be designed with the human user in mind. In automotive design, human factors would include easy-to-read gages, controls at the fingertips, pedals that are well spaced and do not require excessive force to push, steering wheels that are easily turned and ergonomically adjustable, padded seats that do not cause backaches, and so on.

- *Cost.* A product that is aesthetically pleasing, high quality, and user friendly may still fail in the marketplace if it is too costly. There are two types of costs to consider: **initial capital cost** and **life-cycle cost.** The initial capital cost is just the purchase price of the product; the life-cycle cost includes the purchase price but also other costs such as labor, operation, insurance, and maintenance. If it has high maintenance, fuel, and insurance costs, an automobile with a low purchase price actually might be uneconomical compared with a moderately priced automobile. Unfortunately, many consumers only look at the initial capital cost and ignore the life-cycle cost.

- *Safety.* An engineer must always design products that are safe for the end user and the artisans who construct the product. It is impossible to design completely safe products because they would be too costly; therefore, the engineer often must design to industry standards for similar products. Automotive safety standards have been increasing; for example, automobiles have long been equipped with air bags.

- *Operating environment.* The engineer must design the product with the operating environment in mind. What temperature and pressure range will the product experience during storage and use? Is it a corrosive environment? What vibration levels will the product experience? Automobiles must be designed to operate in the Arctic to the Tropics, at sea level to the mountains, in the presence of corrosive road de-icing salt, and on roads filled with potholes.

- *Interface with other systems.* Many products must interface with others: Computers must be compatible with software and printers, televisions must be compatible with broadcast signals. An automobile must be compatible with fuels in common use and must have a turning radius and width compatible with roads.

- *Effect on surroundings.* The creation and use of a product may affect its surroundings adversely. With increasing environmental constraints, products must be designed with lower chemical, noise, and electromagnetic emissions. In the United States, automobiles are equipped with catalytic converters that reduce the air pollution emitted from vehicles. In Germany, automobiles are designed so that when the vehicle reaches the end of its service life, it can be disassembled and the components recycled.

- *Logistics.* Many products require support systems such as electricity, cooling, steam, fuel, and spare parts. Depending on where the product is used, these support systems may or may not be readily available. For example, a product used in outer space has little logistical support, so it must be designed to operate independently from earth. For automobiles, there is a tremendous infrastructure with refueling stations and repair centers widely available, so logistics are less of a problem.

- *Reliability.* A reliable product will always perform its intended function for the required time period and in the environment specified by the user. No product is 100% reliable, although some are close. NASA requires spacecraft components to be highly reliable. This is often accomplished through **redundancy,** that is, using multiple components each with the same capabilities. For example, the space shuttle is controlled by five computers. This provides backup in case any two computers completely fail. Also, if the computers disagree, a "vote" can be taken to settle the dispute. Normally, automobiles are not designed with redundancy, because the consequences of failure are considered less catastrophic.

- *Maintainability.* A product that is maintainable can have the required maintenance performed upon it at the necessary frequency. A satellite is an example of a product that is not easily maintained, because of the difficulty in reaching it. On the other hand, automobiles are highly maintainable, because there is always ready access to a repair shop. *Preventive maintenance* is performed at regular time intervals or when components are nearing failure. Rotating automobile tires at regular mileage intervals and replacing them when they are worn are both examples of preventive maintenance. *Corrective maintenance* is employed after a part fails. Replacing a flat tire with a new tire is an example of corrective maintenance.

- *Serviceability.* If a product can be maintained easily, it is serviceable. For example, if a special tool is required to change an automobile oil filter because it is inaccessible with ordinary tools, it is not serviceable.

- *Availability.* If a product is always ready for use, it is available. If an automobile is frequently in the repair shop and does not operate at temperatures above 80°F or temperatures below 40°F, it is not available a high percentage of the time.

Once the desired properties of the product are identified, it is necessary to weight them, that is, specify their relative importance.

5.1.4 Step 4: Search for Solutions

The search for solutions requires engineers to generate ideas that solve the design problem. Inherently, idea generation is a creative process that cannot be accomplished simply by following a prescribed algorithm. Instead, we offer the following heuristics:

- *Can I eliminate the need?* Suppose you are an automobile designer and your boss defines the need as follows: "The suspension springs are rusting, so I want you to design a coating that protects them." Perhaps you can eliminate the need altogether by specifying nonrusting polymer springs rather than metal springs.

- *Challenge basic assumptions.* The human heart pumps blood through the body in a pulsatile manner, rather than smoothly and continuously. A designer might assume that an artificial mechanical heart must operate in a pulsatile manner also. However, medical studies have shown that after about five days, the human body can adapt to continuous blood flow. A designer who challenges the pulsatile assumption has many more design options available.

- *Be knowledgeable.* In engineering, useful ideas don't spring from nothing; we work from a knowledge base. While searching for solutions, the design engineer must become as knowledgeable about the problem as possible. Information may be obtained from libraries, the Internet, government documents, professional organizations, trade journals, vendor catalogs, and other individuals.

- *Employ analogies.* By using analogies, a design engineer can exploit information from other fields and bring this information to the design problem. Nature is rich with design solutions to problems. For example, in their search for methods to fly a heavier-than-air craft, the Wright brothers employed bird wings as an analogy. Engineers at MIT developed an extremely efficient method to propel boats by making an analogy to fish swimming. Nature is not the only place to find solutions. Many engineers have come before you, so you can make analogies to their solutions to design problems. There are books that catalog engineering solutions. For example, *Pictorial Handbook of Technical Devices* (Schwarz and Grafstein, 1971) has over 5000 illustrations of such technical devices as pumps, gears, bearings, fasteners, machine tools, electronic circuits, knots, bridges, and many others. Thumbing through such a book can be very stimulating when you confront a design problem.

- *Personalize the problem.* For some problems, you can gain insight by imagining yourself shrunken down and literally entering the device being designed. Suppose you are tasked with designing a low-shear pump for transporting fragile polymer solutions. By shrinking yourself down and entering the pump, you can identify the high-shear zones and create a design that eliminates them.

- *Identify the critical parameters.* Many engineering designs have a critical feature that must be overcome for the design to be functional. Perhaps a component must operate at high speeds, or be very reliable, or be made to close tolerances. After these *critical parameters* are identified and addressed, the design will more quickly converge to a workable solution.

- *Switch functions.* Normally, a pump housing is stationary and the internals move. Perhaps you can find an elegant solution to a design problem by switching functions, using stationary internals and a rotating housing.

- *Alter sequence of steps.* Processes involve a prescribed sequence of steps. Perhaps an elegant solution can be developed by altering the sequence. Suppose you are trying to improve the process for making coffee. In the traditional process, first the coffee beans are roasted and then they are ground. This approach works best because traditional grinders function only with roasted beans, which are more friable. But suppose a new grinding technology allows fresh, wet beans to be pulverized. Using this new grinder would allow the traditional steps to be reversed: beans could be pulverized and then roasted. Perhaps the high surface area of the ground beans would allow more flavors to be generated in the roaster, or perhaps the capacity of the roaster would increase when ground beans rather than whole beans are used.

- *Reverse the problem.* Suppose your goal is to develop a lightweight wrench for NASA astronauts. By reversing the problem and thinking of ways you could make the wrench heavier, you will become sensitive to where the weight is and then can think of ways to engineer the weight out.

- *Repeat components or process steps.* Sometimes if one is good, two are better, and three are better yet. Suppose you wish to compress a gas that decomposes if it gets too hot. If the compression is done in a single step, so much compression energy must be added to the gas that it will become too hot and decompose. However, if you break the compression into multiple steps and cool between the steps (intercooling), then you can compress the temperature-sensitive gas to high pressures.
- *Separate functions.* Sometimes an elegant design results when functions are separated. For example, the following functions occur in the piston/cylinder of an automobile engine: air input, fuel input, air/fuel mixing, compression, combustion, expansion, exhaust, lubrication, and wall cooling. Fuel combustion and wall cooling are incompatible functions; perhaps the engine could be improved by performing these functions in separate pieces of equipment.
- *Combine functions.* Sometimes an elegant design results when functions are combined, rather than separated. For example, many industrial processes need both electricity and heat. To supply its energy needs, a plant could convert fuel into electricity in one facility and convert fuel into steam in another. However, these functions may be combined so that fuel is converted into both electricity and steam in a single facility. Such an approach is called *combined heat and power* and is more efficient; the heat that inevitably results from electricity production is used to generate steam rather than being wasted.
- *Use vision.* While you are working on the design, imagine that you have accomplished the goals and are inspecting the final product. What features would this product have if it were built?
- *Employ basic engineering principles.* Throughout your engineering studies, you will learn basic principles that may be applied to your design problem. For example, a basic thermodynamic principle is that unused "driving forces" lead to inefficiencies. Stated another way, when designing a machine or process, every pressure difference, temperature difference, voltage difference, or concentration difference that is not harnessed to make energy will increase total energy expenditures. By reviewing the design and taking steps to minimize these differences, you make the machine or process more efficient.

To help your mind find design solutions, you must put yourself in a physical environment that stimulates creativity. The proper environment will differ from person to person. For some, a busy room filled with people stimulates creativity, whereas others find it distracting. Some people find that everyday activities (driving, gardening) stimulate creativity, for these activities do not require focused attention but allow the mind to wander freely and explore creative solutions. Sometimes while trying to find a solution, you may hit a "brick wall" and make no progress. In this situation, it pays to take a break and do something completely different, like play music or sports, or even sleep. When you return to the problem, you may find that your subconscious solved the problem while you were engaged elsewhere. Sometimes the biggest aid to creativity is simply dogged perseverance; by focusing on the problem for an extended period of time, you completely immerse yourself in it and increase your chances of finding a solution.

Although an individual can certainly generate creative ideas, it helps to work with others. The interaction stimulates ideas that no lone individual could find. The following techniques have been developed as formats to help people work together to solve problems:

Brainstorming

Brainstorming is such an important tool in the creative part of engineering that it's appropriate to say a bit more about it. The magic of brainstorming falls into two parts.

First, there's something about being with a group of creative people focused on solving a problem that helps shake the ideas from your head. Some of our best ideas have "fallen from our mouths" during these sessions. The atmosphere in a well-formed group somehow frees you to pull things together that otherwise, by yourself, you would probably not have connected or it would have taken much longer.

The second bit of magic during brainstorming sessions comes from the interaction of the group building on each other's ideas. A couple of (probably apocryphal) stories should help to clarify why ideas thrown out before the group should not be dismissed or belittled at the time. During naval operations, minesweepers would drag nets through an area of the sea that was to be cleared. Usually, the mines would be caught in the nets between two minesweepers and were exploded harmlessly by shooting them with machine guns. Sometimes, however, the mines would become entangled in the nets close to the minesweepers themselves and, if exploded, would damage or sink the ships. Several approaches had been tried including cutting the nets or launching small boats to try to drag the mine away, but none worked well. A group of navy personnel was gathered to formulate a general strategy. During the brainstorming session, one of the group suggested that everyone go on the deck, purse their lips, and blow the mine out to sea where it could be shot as normal. As the smiles died away, another in the group exclaimed, "We could use the deck fire hoses to push the mines away!!" Sure enough, this became the standard procedure.

Another story involved removing prime logs from a forest without damaging the surrounding forest with roads. After many vain attempts to design roads that would minimally damage the mountainsides, one of the group suggested that the only way was to get little green men from Mars to come in their space ships and haul the logs away. Another suggested that he didn't know about space ships, but he bet that helicopters would work. And indeed they do.

Much has been written about brainstorming and group creativity; explore this area as you develop your engineering skills. Our final thought for you is that, as with many things, your brainstorming skills will improve with practice.

- *Brainstorming.* The group (generally 6 to 12) has a leader that sets the tone for free expression. All members reveal their ideas to the group as soon as they form them. The most important rule is not to belittle any ideas or rate them. Often, seemingly silly ideas lead to valuable solutions. A person is designated to record the ideas as they are generated.
- *Nominal Group Technique.* In this technique, the leader assembles the group and poses the problem. Each group member independently works on the problem and writes ideas on paper. Then, when the group members have stopped generating ideas, each group member sequentially explains one idea to the entire group. Each idea is recorded on a whiteboard or flipchart for all members to see. Although discussion is permitted for clarification purposes, no critiques are allowed. Later, when the ideas are all identified, the group members may be asked to rate the ideas. If a few personalities in the group are dominant and tend to sway discussions, this technique reduces the potential for "groupthink," which might occur with brainstorming.
- *Delphi Technique.* This technique is similar to the previous technique except that the members of the group are intentionally separated. The leader e-mails each group member the problem statement. The group members then return their solutions by e-mail. The leader pools the responses and asks the group members to rate the ideas. Those ideas receiving low ratings can be clarified by the originator, if desired.

The Color TV War

The foundations for television were laid in 1883 when Germany's Paul Nipkow developed a spinning disk with spiral apertures that broke an image into a series of scanned lines. In 1889, Russia's Polumordvinov conceptualized a system in which a Nipkow disk was combined with three color filters that would break the image into the three primary colors (red, green, blue). His efforts, and those of other inventors (Adamian, von Jaworski, and Frankenstein), never resulted in a workable system.

In 1928, a workable spinning-disk color television was demonstrated in England. A year later, Bell Labs demonstrated a similar system in the United States. However, during the 1930s, most development efforts focused on black-and-white television.

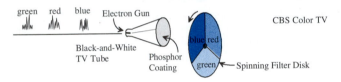

Scanned Field — Nipkow spinning disk

The image is scanned once per revolution.

Aperture

In 1940, Peter Goldmark of CBS developed a color television that combined a spinning multicolor disk with a black-and-white television tube in which an electron gun "painted" the image on a phosphor coating. In Goldmark's system, the three primary colors were broadcast *sequentially*. When the blue filter aligned with the TV tube, the blue signal reached the TV tube and the eye would see only the blue information. The other primary colors were received by the eye in a similar manner. Because the disk was spinning rapidly, the eye integrated the three primary colors into a single image.

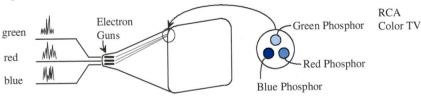

green   red   blue   Electron Gun   CBS Color TV

Black-and-White TV Tube   Phosphor Coating   blue red green   Spinning Filter Disk

In July 1941, commercial black-and-white broadcasting began to an audience of a few thousand, but in December, World War II intervened and halted progress on both black-and-white and color television. After the war, in 1946, RCA sold 10,000 black-and-white sets at a cost of about $385 each. In the same year, CBS demonstrated Goldmark's television to the Federal Communications Commission (FCC), seeking approval for this to be the standard color television. RCA waged a campaign against adopting the CBS television because its signal would be incompatible with black-and-white sets already being sold. To make the black-and-white sets compatible would require the consumer to spend about $100 for a special converter.

green   Electron Guns   Green Phosphor   RCA Color TV

red   Red Phosphor

blue   Blue Phosphor

In 1946, RCA was developing its own all-electronic color television in which the three primary colors were broadcast *simultaneously*. Their television tube had three electron guns, one for each color. The viewing surface of the tube had three phosphors; each would glow a separate primary color when struck by the electron

beam. Further, through clever engineering, their signal would be completely compatible with black-and-white television.

In 1950, the CBS and RCA systems were tested side by side. The *Variety* headline announced "RCA Lays Colored Egg" and proclaimed the failure of RCA compared to CBS. The FCC authorized CBS to begin producing its sets.

RCA countered by stepping up production of black-and-white sets, which were incompatible with the CBS color broadcast signal. According to RCA's David Sarnoff, "Every set we get out there makes it that much tougher on CBS." Sarnoff increased RCA's efforts to perfect its system by requiring employees to work 18 hours per day, including weekends. Rewards of thousands of dollars were offered for key developments. Sarnoff tried, unsuccessfully, to get the courts to block the FCC ruling.

In 1951, CBS broadcast the official color premier, a one-hour Ed Sullivan show, to a few dozen color sets. In contrast, there were now 12 million black-and-white sets in the United States.

RCA's development efforts paid off. In 1951, a few days after the CBS premier broadcast, RCA demonstrated its new color television with a 20-minute program. The press responded enthusiastically to the demonstration. Clearly, the RCA system was the way of the future. In 1953, CBS abandoned its efforts in color television, and the FCC officially adopted the RCA system.

In 1954, the first RCA color sets were offered for $1000 each (a quarter of the average annual salary). Only 5000 sets were sold, out of a projected 75,000. Further, the color sets were failure prone; twice as many service calls were required for the few thousand color sets than all the millions of black-and-white sets put together. Over the years, prices were cut but sales were sluggish. *Time* magazine declared color television "the most resounding industrial flop of 1956."

By 1959, RCA had spent $130 million developing and marketing color television and still had not made a profit. However, in 1960, they acquired the Walt Disney Show from ABC. NBC announced it would broadcast the show in color as "Walt Disney's Wonderful World of Color." By 1960, RCA made its first profit in color television.

Interestingly, although the CBS spinning-disk system was not successful in the consumer market, it was successful with NASA. During the Apollo moon mission, NASA needed a color TV camera that was compact, lightweight, robust, energy-efficient, simple to use, and sensitive to low light levels. At the time, the RCA system could not meet the objectives, but the CBS system could. From the moon, the CBS sequential signal was broadcast to earth where it was converted to the RCA simultaneous signal so it could be viewed by millions of people all over the world.

Adapted from: D. E. Fisher and M. J. Fisher, "The Color War," *American Heritage of Invention & Technology* 12, no. 3 (1997), pp. 8–18; and S. Lebar, "The Color War Goes to the Moon," *American Heritage of Invention & Technology* 13, no. 1 (1997), pp. 52–54.

5.1.5 Feasibility Study

A feasibility study is meant to quickly eliminate ideas without consuming much engineering time. The goal is to see the big picture and address the most important features of the problem.

Step 5: Analyze Each Potential Solution.

During the feasibility study, when first analyzing each potential solution, an engineering team may use simple calculations to characterize each design. Alternatively, the team may use simple heuristics (e.g., number of parts, number of process steps, number of high-precision components, simplicity of components, logistical complexity, or need for exotic materials) to analyze different design options.

Step 6: Choose the Best Solution(s).

Although engineers would like to investigate each candidate technology in great detail, this is rarely possible. Because we do not have infinite time to work on a problem, it is necessary to eliminate some technologies based upon the results of the feasibility study. Depending upon the methods used to analyze each design option, engineers develop a rating scheme to choose the best option(s). For example, designs with fewer numbers of parts tend to be less expensive and more reliable. So the team may select the designs that have the fewer number of parts.

Step 7: Document the Solution(s).

After the engineers have searched for a solution and made some choices, they should document their results in written form. The report should define the problem, identify criteria for success, describe and analyze the various options, describe the rating scheme used to evaluate the various options, and recommend the best option(s). Graphics and good writing are essential to the report. This report is distributed to team members so everyone is working from the "same sheet of music." Before the report is released, all members of the team must concur with it.

Step 8: Communicate the Solution(s) to Management.

The engineers must communicate their results to management by sending them a copy of the written report developed in Step 7. In addition, they may wish to discuss the results in person, by phone, or in a formal oral presentation.

5.1.6 Preliminary Design

The purpose of the feasibility study was to do a "quick and dirty" survey to determine if more effort is warranted. Provided the outcome was positive, the engineers now engage in a preliminary design that is more detailed than the feasibility study.

Step 5: Analyze Each Potential Solution.

During the preliminary design, the engineers will use detailed calculations to analyze the promising designs that emerged from the feasibility study. Analysis of the design options

TABLE 5.1
Evaluation matrix

Property	Weighting Factor	Option A		Option B		Option C	
		Score	Weighted Score	Score	Weighted Score	Score	Weighted Score
Property 1	α	A1	$\alpha \times$ A1	B1	$\alpha \times$ B1	C1	$\alpha \times$ C1
Property 2	β	A2	$\beta \times$ A2	B2	$\beta \times$ B2	C2	$\beta \times$ C2
Property 3	γ	A3	$\gamma \times$ A3	B3	$\gamma \times$ B3	C3	$\gamma \times$ C3
Total			Σ above		Σ above		Σ above

may rely heavily on engineering courses such as thermodynamics, statics, strength of materials, circuit analysis, heat transfer, fluid flow, and others.

Step 6: Choose the Best Solution(s).

The preliminary design analysis provides us additional information with which to identify the best solution(s). One approach to identifying the best solutions is to list the advantages and disadvantages of each option. Another approach uses an **evaluation matrix** (Table 5.1), in which the desired properties and their relative importance (i.e., weighting factors) are listed on the left. Each option has a separate column in which a score is given for each property. The scores multiplied by the weighting factors are added for each option. The option with the highest total weighted score is selected.

Steps 7 and 8: Document the Solution(s) and Communicate with Management.

The results of the preliminary design should be documented in a report. First the report should be circulated among the design team to get everyone's concurrence; then it is sent to management. This report contains the same topics as the feasibility report, but with more details. In addition to the written report, the results of the preliminary design are presented to management in a formal oral report that allows them to ask questions to determine if the project is worth continuing. Normally, a single solution will emerge from the preliminary design phase.

5.1.7 Detailed Design

When a project successfully passes through the preliminary design phase, it enters the detailed design phase. The detailed design involves a very large team that works on *the* solution that emerged from the preliminary design phase.

Steps 5 through 8 of the Detailed Design.

Each and every component of the solution must now be specified in great detail. Issues such as materials, dimensions, tolerances, and processing steps all must be documented in detailed drawings and reports that allow artisans to construct a prototype. For some complex projects, such as airplanes, the detailed drawings and reports literally weigh hundreds or thousands of pounds and require a truck to transport them.

5.1.8 Step 9: Construct Solution

Typically, a prototype will be constructed from the documents produced in the detailed design. If the prototype testing goes well, the design will be finalized. A production facility must be built to construct the product to be marketed to the consumer. Many people are involved in this phase. Material suppliers must be identified and contracts written. Personnel to operate the production facility must be hired. Large-scale processes must be refined, because the prototype was constructed in a small shop. Marketing strategies must be finalized and financing arranged. Owner's manuals must be written, to go with the product. Spare parts must be manufactured and stocked to provide future customer support.

5.1.9 Step 10: Verify and Evaluate

The engineer's job is not complete once the first product is made in a large-scale production facility. Product samples should now be taken from the production facility to determine that the product is meeting the design specifications. If there are problems in the production facility, they must be corrected.

The engineer now should be thinking about the next generation of products. Because technology changes rapidly, the engineer should be thinking about improved manufacturing techniques, substituting new materials with superior properties, or improving the design to meet the latest market requirements. If management approves the next generation of products, the engineer returns to Step 1, where the design problem is defined.

5.2 FIRST DESIGN EXAMPLE: IMPROVED PAPER CLIP

To illustrate the steps of the engineering design method, we consider the design of an ordinary object: the paper clip.

Step 1: Identify the Need and Define the Problem. A salesperson working for Office Supply, Inc. (OSI) returned from a recent sales call to a hospital. The nurses informed him that a patient received improper treatment and nearly died because a critical document was lost from a stack of papers fastened with ordinary paper clips. Apparently, a nurse was carrying a stack of 20 bound medical records and dropped them. The paper clips failed, thus scattering the papers. In the resulting confusion, a critical document was placed in the wrong file.

The salesperson approaches OSI management, requesting that a new paper clip be developed for holding critical documents, such as medical records. Management likes the idea; high-end products often command a greater profit margin.

The salesperson has identified the need for a highly reliable paper clip. Management has defined the problem as, "Develop a highly reliable paper clip for use in high-end applications, such as binding medical documents."

Step 2: Assemble Design Team. To design an improved paper clip, management assembles a design team composed of the following individuals: mechanical engineer for specifying the shape of the clip, manufacturing engineer for specifying the equipment to manufacture the clip, and marketing expert to determine how to market the clip to high-end users.

Step 3: Identify Constraints and Criteria for Success. The marketing expert determines that to be successful, the maximum acceptable price is $1.00 per clip. The clip

TABLE 5.2
Properties of improved paper clip

Property	Weight	Justification
Reliability	5	The paper clip is intended for critical documents, so reliability is essential.
Compact design	3	Although a compact design is desirable, it is not critical.
Weight	3	Although light weight is desirable, it is not critical.
Convenience	5	The marketing expert indicated that no matter how good the new paper clip is, if it is inconvenient to use, it will not sell.
Low cost	1	This paper clip will be marketed to high-end customers, so cost is not as much of a factor.

system must be capable of binding anywhere from 2 to 200 pages. The pages cannot separate if the papers are dropped from a height of 10 feet or lower.

The design team decides that the paper clip should have the properties shown in Table 5.2. The relative importance of each property is indicated with a weight; justifications are given for each assigned weight.

Step 4: Search for Solutions.   The design team did some research on paper clips and found an article on the subject.[1] The article states that before there were paper clips, papers were joined with a straight pin. This technology had obvious problems, such as being limited to only a few sheets of paper, sticking the reader with the sharp point, and grabbing extraneous papers. In the mid-19th century, clothespins and elaborate wooden clips were used to bind papers. By the late-19th century, machines for bending wire into paper clips became available, and a myriad of shapes were developed (Figure 5.2). Near the turn of the century, the standard "Gem" paper clip became available from the British company Gem, Ltd.

Commonly available paper clips all rely on the springiness of the wire to hold the papers together. Other options should be identified for holding the paper more securely. Analogous situations should be explored.

From the Petroski article, the design team learned that clothespins have been used to hold papers. The design team discussed this possibility and developed Options 1 and 2 (Figure 5.3). However, rather than rely on a spring to clamp the papers, they proposed two more secure possibilities. Option 1 employs a screw that, when extended, tightly clamps the papers. Option 2 uses a wedge to clamp the papers.

Options 1 and 2 are bulky, so the design team did some brainstorming and came up with Options 3 and 4. Both these design options are analogous to a vise or clamp. The sandwiched metal strip pivots as the screw is rotated or the wedge is inserted. Options 5 and 6 are very similar to Options 3 and 4, except that the sandwiched metal strip slides along a guide instead of pivoting.

Options 1, 3, and 5 require a screwdriver to adjust the screw; Options 2, 4, and 6 require a wedge that could be lost. Desiring a self-contained paper clip, the team devised

1. H. Petroski, "The Evolution of Artifacts," *American Scientist* 80 (1992), pp. 416–420.

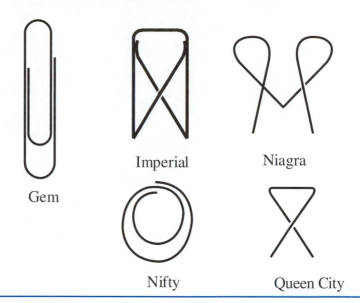

FIGURE 5.2
Examples of wire paper clips.

Option 7 (Figure 5.4). The paper is inserted into the paper clamp, which has a square hole on the top. The retractable pin is threaded at the top and square at the bottom, where it fits into the square hole of the paper clamp; the square hole prevents the retractable pin from rotating. The threaded portion of the retractable pin fits into the threaded hole in the thumbwheel. The thumbwheel is captured by the thumbwheel retainers, which allow the thumbwheel to rotate, but prevent movement in the axial direction. As the thumbwheel rotates, it causes the retractable pin to move downward and clamp the papers. A variety of sizes could be manufactured to bind different numbers of papers.

Step 5: (Feasibility Study) Analyze Each Potential Solution. The design team decided to use an evaluation matrix (Table 5.3) to help select the best option. They rated the five properties of each option as follows:

- <u>Reliability</u>. Options 1, 3, 5, and 7 use a screw that is unlikely to fail; therefore, these options scored the highest points. Options 2, 4, and 6 use a wedge that might fall out; hence, these options scored fewer points.
- <u>Compact design</u>. Options 1 and 2 are very large, so they scored poorly. Also, the options with a wedge scored poorly because the wedge takes up space. The thumbwheel on Option 7 requires some space, so this option scored poorly also. Of all the options, Options 3 and 5 were the most compact; even so, they were not as compact as traditional paper clips, so they scored only 3 points.
- <u>Weight</u>. The weight scores are similar to the compactness scores, except for Option 7, where some of the components could be made from plastic to reduce weight.
- <u>Convenience</u>. Options 1, 3, and 5 require a screwdriver, which is inconvenient. Options 2, 4, and 6 require the wedge, which could easily be lost. Only Option 7 was deemed convenient, so it got maximum points.
- <u>Cost</u>. Option 7 is the most complex, so it has the highest cost. The other options are less expensive than Option 7 but still costly compared to traditional paper clips.

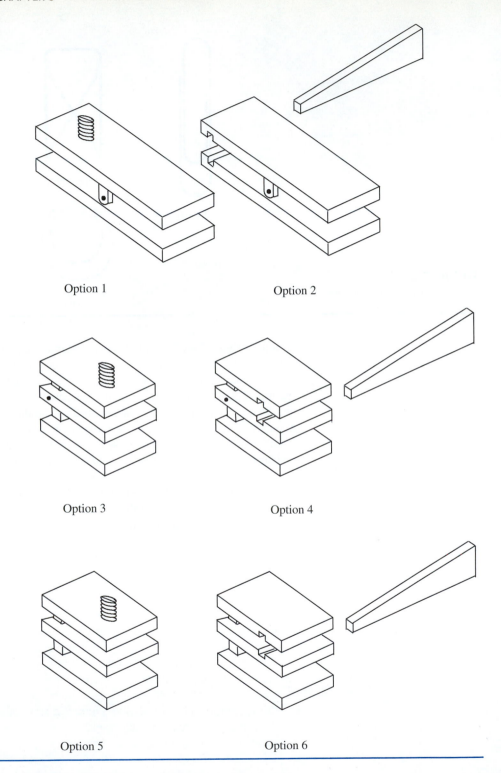

Option 1

Option 2

Option 3

Option 4

Option 5

Option 6

FIGURE 5.3
Six options for an
improved paper clip.

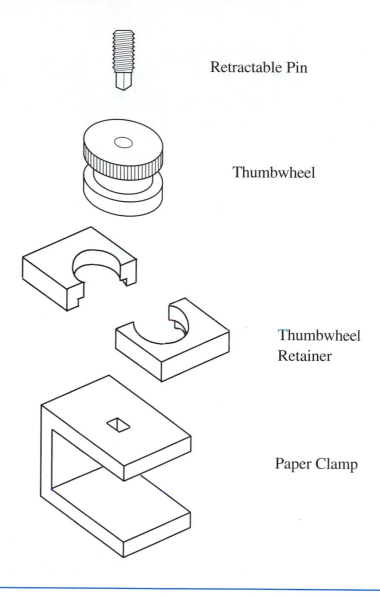

Retractable Pin

Thumbwheel

Thumbwheel
Retainer

Paper Clamp

FIGURE 5.4
Option 7 for an
improved paper clip.

Step 6: (Feasibility Study) Choose the Best Solution(s).   Based on the total weighted scores shown in Table 5.3, Option 7 is the best by a wide margin.

Step 7: (Feasibility Study) Document the Solution(s).   The design team prepares a written report documenting the various options and their evaluation.

Step 8: (Feasibility Study) Communicate the Solution(s) to Management.   The design team makes an oral presentation to management to explain the options and why they think Option 7 is the best. They submitted their written report to management one

TABLE 5.3
Evaluation matrix for improved paper clip

Property	Weight W	Option 1 S	Option 1 S × W	Option 2 S	Option 2 S × W	Option 3 S	Option 3 S × W	Option 4 S	Option 4 S × W	Option 5 S	Option 5 S × W	Option 6 S	Option 6 S × W	Option 7 S	Option 7 S × W
Reliability	5	5	25	3	15	5	25	3	15	5	25	3	15	5	25
Compact design	3	1	3	1	3	3	9	2	6	3	9	2	6	2	6
Weight	3	1	3	1	3	3	9	3	9	3	9	3	9	3	9
Convenience	5	1	5	1	5	1	5	1	5	1	5	1	5	5	25
Low cost	1	4	4	4	4	3	3	3	3	3	3	3	3	5	5
Total			40		30		51		38		51		38		70

1 = Poor, 5 = Excellent

week prior to the oral presentation so management had time to review the options and think of good questions.

After the oral presentation, management approves the project for continued support.

Steps 5 through 8 of the Preliminary Design.

The engineering team performs some detailed calculations to determine the required thickness of the metal in the paper clamp. They decide it will be constructed from ordinary carbon steel that is electroplated to resist corrosion. The retractable pin will also be constructed of electroplated metal. To save weight and cost, the thumbwheel will be plastic, with a metal insert for the threads. The thumbwheel retainers will also be plastic.

The machine shop makes some prototype clips and tests them to work satisfactorily when papers are dropped from a height of 10 feet.

To document their design, the team prepares engineering drawings and written reports. After a design review with management, the project receives approval to continue.

Steps 5 through 8 of the Detailed Design.

Although the prototype performed successfully, still there are many issues to address, primarily related to manufacturability. To meet the cost objective of $1 per paper clip, it is critical to identify manufacturing methods that can produce and assemble the components cost-effectively. Once the manufacturing machines and methods are identified, the final engineering drawings of the paper clip can be prepared. The drawings address key issues such as material suppliers and manufacturing tolerances for each component.

Step 9: Construct the Solution.

The machinery to construct the paper clips is installed and paper clips are sold to high-end customers.

Step 10: Verify and Evaluate.

The paper clips must be constantly sampled and inspected to ensure that the manufacturing equipment continues to make high-quality products. Even though the process is operating successfully, the engineers should always be looking for improved manufacturing methods to reduce costs and prevent loss of market share to competitors.

5.3 SECOND DESIGN EXAMPLE: ROBOTIC HAND FOR SPACE SHUTTLE

To continue our familiarization with the engineering design method, we explore the design of a robotic hand used to build the International Space Station.

Step 1: Identify the Need and Define the Problem.　The U.S. Congress authorized funding for NASA to construct the International Space Station, an inhabited satellite providing permanent manned presence in near-earth orbit. Much of the hardware was provided by other nations, so the project was truly international in scope. The purpose of the International Space Station is to provide a microgravity environment for scientific and engineering investigations. Also, it allows physicians to study the long-term effects of weightlessness on human physiology.

Although the hardware is manufactured on earth, it must be assembled in space. Some components are very large and cannot be manipulated directly by free-floating astronauts wearing protective space suits; therefore, large items are manipulated by a robotic hand. The robotic hand is attached to the end of a robotic arm mounted on the space shuttle. To assemble a large component, the operator manipulates the object using the robotic arm and hand, which are governed by remote controls located inside the space shuttle.

Although the space shuttle already had a robotic arm and hand, the hand was too small to manipulate the large components of the space station; therefore, your company has been hired to design and construct a large robotic hand to replace the small hand.

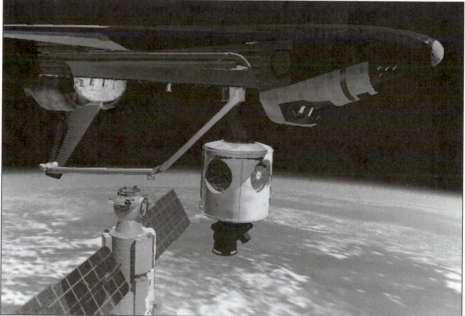

The space shuttle can manipulate components of the International Space Station using its robotic arm.

Photo courtesy of NASA.

TABLE 5.4
Evaluation matrix for robotic hand actuators

Property	Weighting Factor	Explanation
Soft grip	10	Essential so that space station components are not crushed.
Failure does not damage space station or shuttle	10	Failure of the robotic hand should not lead to other failures, risking lives.
Energy efficient	5	The robotic hand will not be utilized frequently, so energy efficiency is not critical.
Mass	5	If extra mass is needed for functionality, the extra cost is affordable.
Reliability	9	A failure could cause a mission to be "scrubbed," which is very expensive.
Operates over full temperature range	5	The temperature changes frequently, so it can be used when the temperature is acceptable.
Compatible with vacuum	7	Incompatibilities are undesirable, but acceptable if they do not damage space station or shuttle.
Cost	6	Cost overruns are undesirable, but are preferable to developing inferior equipment.
Time schedule	6	Time overruns are undesirable, but are preferable to developing inferior equipment.

Step 2: Assemble Design Team.   To satisfy the NASA contract, your management has formed a design team composed of individuals with the following expertise: project management, mechanical engineering, electrical engineering, control engineering, and manufacturing.

Step 3: Identify Constraints and Criteria for Success.   To meet the contract, the robotic hand must be constructed within 1 year at a cost of $10 million. Because it costs $10,000 to lift 1 pound into space, the contract specifies that the hand can have a mass no larger than 100 lb_m. Because the robotic hand will be used in either full sun or complete darkness, it must operate in a temperature range from -150 to $+100°C$. It must function in a complete vacuum and will be subjected to micrometeor showers, which are frequent in space. Because it is difficult to provide service in space, the robotic hand must be very reliable. If it fails, it cannot damage the space shuttle or space station. The robotic hand must have a "soft grip," meaning it cannot crush the space station components while assembling them. Because energy is precious in space, the robotic hand must be energy efficient.

The design team discusses the relative importance of each of these features. They establish weighting factors for each feature, with 10 indicating the most weight and 1 the least weight. Table 5.4 shows the weighting factors and an explanation for each.

Step 4: Search for Solutions.   The design team meets in a brainstorming session and discusses the robotic hand. Because it will be used to manipulate large objects, fine dexterity is not required. Team members decide that it can be more like a lobster claw than a human hand. After discussing a number of ideas, they sketch the two concepts in Figure 5.5. Option 1 has an **actuator** that pulls on cables that manipulate two spring-loaded fingers. The finger tips are rubber, thus providing a soft grip. Option 2 has two stationary fingers and one actuated opposable thumb.

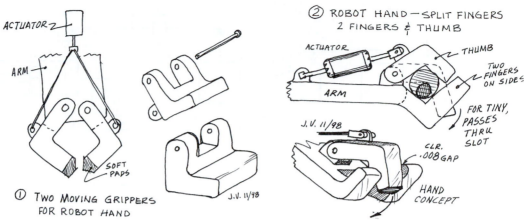

FIGURE 5.5

Sketches of two robotic hands.
(Drawings kindly provided by Gerald Vinson.)

Each of the robotic hands requires an actuator. The design team develops the three options illustrated in Figure 5.6. Notice that numbers are used to indicate component parts, rather than words. This numbering technique is often used in patents so that the drawings do not become cluttered. Also, it allows for completely unambiguous labeling of components. For the same reasons, we use this numbering technique here.

Option A uses pneumatic pressure as the actuator. Low-pressure gas is stored in tank 1. Compressor 2 pressurizes the low-pressure gas so it can flow into high-pressure tank 3. Pressure regulator 4 adjusts the downstream gas pressure; a high pressure gives a firm grip and a low pressure gives a soft grip. Double-acting piston 12 actuates push rod 11. By opening valves 5 and 8 and closing valves 6 and 7, the pressure in chamber 13 is greater than in chamber 14, so push rod 11 extends. Similarly, by closing valves 5 and 8 and opening valves 6 and 7, the pressure in chamber 13 is less than chamber 14, so push rod 11 retracts. Relief valves 9 and 10 protect chambers 13 and 14 in case of overpressure.

Option B uses hydraulic pressure as the actuator. Low-pressure hydraulic fluid is stored in accumulator 21, a tank with a gas-filled bladder. Pump 22 pressurizes the low-pressure hydraulic fluid so it can flow into high-pressure accumulator 23. Pressure regulator 24 adjusts the downstream hydraulic pressure; a high pressure gives a firm grip and a low pressure gives a soft grip. Double-acting piston 32 actuates push rod 31. By opening valves 25 and 28 and closing valves 26 and 27, the pressure in chamber 33 is made greater than in chamber 34, so push rod 31 extends. Similarly, by closing valves 25 and 28 and opening valves 26 and 27, the pressure in chamber 33 is made less than in chamber 34, so push rod 31 retracts.

Option C uses an electric servomotor 41 as the actuator. In a servomotor, the number of shaft rotations can be controlled. The threaded shaft 42 has a nut 43 that can travel along the length of shaft. Extensions 44 connect to piston 45 located inside of cylinder 47. When nut 43 travels rightward, spring 46 pushes against baseplate 48 to extend push rod 51. Similarly, when nut 43 travels leftward, spring 46 pulls against baseplate 48 to retract push rod 51. When grasping an object, the more spring 46 is compressed, the tighter the grip.

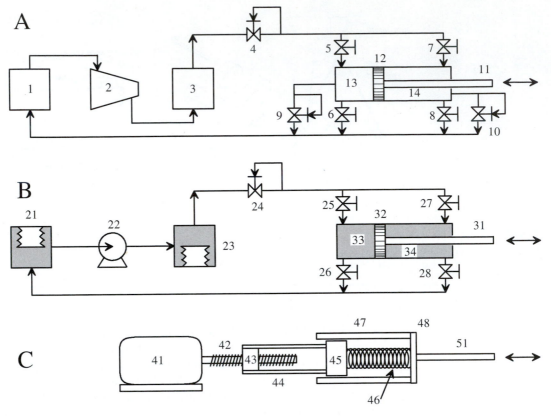

FIGURE 5.6
Schematic of actuators. Options: A = pneumatic, B = hydraulic, and C = electric motor.

Step 5: (Feasibility Study) Analyze Each Potential Solution.   To analyze which of the two robotic hands is better, the engineers list the advantages and disadvantages of each.

	Disadvantages	**Advantages**
Option 1	Difficult to grasp both large- and small-diameter objects.	Spring load on the fingers sets the maximum squeezing pressure. It can be adjusted to prevent damage to space station components.
Option 2	Actuator could squeeze too hard and damage space station components.	Can grasp both large- and small-diameter objects.

To analyze which of the actuators is best, the design team creates an evaluation matrix (Table 5.5). Here are explanations of the scores:

- Soft grip. All designs have the soft grip feature.
- Failure does not damage space station or shuttle. If a micrometeor ruptures the pneumatic lines, the robotic hand could flail around because the exhausting gas behaves like

TABLE 5.5
Evaluation matrix for robotic hand actuators

Property	Weighting Factor	Option A		Option B		Option C	
		Score	Weighted Score	Score	Weighted Score	Score	Weighted Score
Soft grip	10	10	100	10	100	10	100
Failure does not damage space station or shuttle	10	8	80	1	10	10	100
Energy efficient	5	1	5	5	25	10	50
Mass	5	5	25	1	5	10	50
Reliability	9	3	27	7	63	10	90
Operates over full temperature range	5	8	40	1	5	10	50
Compatible with vacuum	7	4	28	6	42	10	70
Cost	6	10	60	10	60	10	60
Time schedule	6	10	60	10	60	10	60
Total			425		370		630

1 = Poor, 10 = Excellent

a miniature rocket motor. However, the robotic arm is fairly rigid, so movement should be limited. If a micrometeor ruptures the hydraulic lines, hydraulic fluid could be sprayed onto the space station or shuttle and may damage critical components. None of these failures can occur with the electric motor.

- _Energy efficient_. Because the gas is compressible, large gas volumes must be compressed to achieve a given pressure, so much energy is needed. The pressure regulator in the hydraulic line causes energy losses. No such losses occur with the electric motor.
- _Mass_. The gas tanks and compressor are fairly heavy. Also, the hydraulic fluid is heavy. The electric motor eliminates these masses.
- _Reliability_. The pneumatic and hydraulic options are complex because of the many valves and pressure regulators, and therefore violate the KISS principle. In contrast, the electric motor is the simplest design.
- _Operates over full temperature range_. The gas pressure in the pneumatic system will fluctuate as the temperature changes, making control difficult. The viscosity of the hydraulic fluid will vary greatly with temperature. None of these problems occur with the electric motor.
- _Compatible with vacuum_. Both the pneumatic and hydraulic systems have a critical seal around the push rod. An imperfect seal means fluid will leak and must be replaced. This problem does not exist with the electric motor.
- _Cost_. All designs have a comparable cost.
- _Time schedule_. All designs can meet the design schedule.

Step 6: (Feasibility Study) Choose the Best Solution(s).
The design team discusses the advantages and disadvantages of each robotic hand. Because space station components come in many sizes, they believe Option 2 is the better design.

For the actuator, according to the evaluation matrix in Table 5.5, Option C is clearly the best design.

Step 7: (Feasibility Study) Document the Solution(s).   The design team prepares a written report documenting the various options and their evaluation.

Step 8: (Feasibility Study) Communicate the Solution(s) to Management.   To allow time to review the written report, the design team gives it to management one week prior to the design review meeting. At the meeting, the team verbally presents their work and answers questions raised by management. Management agrees with the design team's conclusions and authorizes them to proceed with the preliminary design.

Steps 5 through 8 of the Preliminary Design.   The engineering team prepares a preliminary design of robotic hand Option 2 and actuator Option C. They address such issues as identifying potential vendors for the servomotor 41, materials of construction for various components, lubrication for the threaded rod 42 and nut 43, methods for joining the spring 46 to the piston 45 and baseplate 48, methods to control the servomotor 41, sensors to determine the amount of compression in the spring 46, and methods of machining or casting the fingers. When addressing each issue, multiple options are available, so they must choose the best option using an evaluation matrix, or a table listing the advantages and disadvantages of each option. Again, they must document their decisions in a written report and defend their choices to management. Once management approves the preliminary design, the design team can proceed with the final design.

Steps 5 through 8 of the Detailed Design.   During the detailed design, team members draw each component, indicating dimensions, tolerances, and materials of construction. They also prepare assembly drawings and instructions. Figure 5.7 shows an example of a drawing of the assembled components. These drawings must be approved by management before construction can begin.

Step 9: Construct the Solution.   While the robotic hand is being constructed, the design team answers questions from the artisans (e.g., machinists, electronic technicians). Because the artisans have more hands-on experience than the engineers, they provide many useful suggestions during construction. During this phase, the engineers must be sensitive to problems that may arise and modify the design if necessary.

Step 10: Verify and Evaluate.   Once the robotic hand construction is complete, it must be tested to ensure that it meets the design specifications and requirements. If testing shows there are deficiencies, the team must modify the design and the artisans must construct replacement components. Modifications at this stage are extremely expensive because they cause delays; however, if the hardware is deficient, these delays may be unavoidable.

5.4   SUMMARY

An engineer is a person who applies science, mathematics, and economics to meet the needs of humankind; they meet these needs by implementing the engineering design method to design new products, services, or processes.

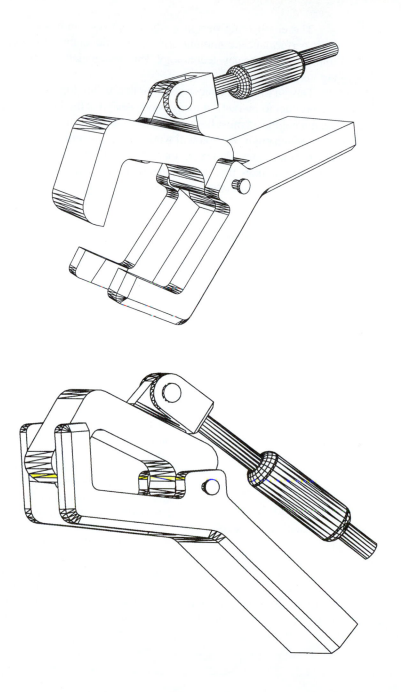

FIGURE 5.7
Example of a detailed drawing. (Drawings kindly provided by Gerald Vinson.)

The engineering design method involves synthesizing new technologies, analyzing their performance, communicating within the engineering team and to management, and implementing the new technology. This design method consists of 10 steps, many of which are iterative.

Design involves conceptualizing new technologies to meet human needs and bringing the product to market. In many ways, design is the heart of engineering. To be successful, design engineers must have excellent creative, technical, organizational, and people skills. As a design engineer, you will work with engineers from many disciplines as well as many other professions (lawyers, marketers, advertisers, financiers, distributors, politicians, etc.). It is unlikely that your formal schooling will teach you everything you need to know, so you must continue to learn after graduation.

Further Readings

Beakley, G. C., and H. W. Leach. *Engineering: An Introduction to a Creative Profession,* 4th ed. New York: Macmillan, 1982.
Norman, D. A. *The Design of Everyday Things.* New York: Basic Books, 2002.
Schwarz, O. B., and P. Grafstein. *Pictorial Handbook of Technical Devices.* New York: Chemical Publishing, 1971.

PROBLEMS

5.1 Identify an everyday annoyance and design an engineering solution to the problem.

5.2 Pick one of the following household items and list as many creative uses as possible:

(a) coat hanger

(b) spoon

(c) toothpick

(d) piece of paper

5.3 Design a system for filling an automobile gas tank that does not emit gasoline vapors into the air.

5.4 How could you keep raindrops from blocking the view through a windshield without using squeaky windshield wipers? List five ideas.

5.5 Design a system that solves the flat-tire problem by (1) preventing flat tires; (2) quickly repairing flat tires; or (3) quickly changing the flat tire with the strength of a 10-year-old child.

5.6 A raw egg is dropped from the third floor of a building to a catcher located on the ground floor. Design a catcher that does not break the egg.

5.7 Design a system that places radioactive tennis balls into a safe bucket without human touch. The system must be operated by humans located on a balcony overlooking a ground floor, which is where the radioactive tennis balls are located.

5.8 Design a system to take soil samples from a 1-meter depth on Mars. The system can weigh no more than 10 kg.

5.9 Design a method that prevents the Leaning Tower of Pisa from falling over.

5.10 Design a system that turns the pages of a book for a disabled person who does not have use of arms.

5.11 Design an inexpensive system that dispenses and opens plastic bags that hold fruit or vegetables for customers shopping at a grocery store.

Glossary

actuator A mechanism that uses pneumatic, hydraulic, or electrical signals to activate equipment.

concurrent engineering A design approach in which specialists work together right at the beginning of a project.

Delphi technique A technique in which the members of a group are intentionally separated and given the same problem. They return their solutions to the leader, the solutions are pooled, and the group members rate the ideas.

detailed design The step where highly detailed drawings and specifications are prepared for the best design option.

evaluation matrix A mathematical evaluation that identifies the best solution by weighting desired properties on the basis of their relative importance.

feasibility study The step of the engineering design method where ideas are roughed out.

global optimum The best condition found without constraints.

initial capital cost The purchase price of a product.

life-cycle cost The purchase price and additional costs such as labor, operation, insurance, and maintenance.

local optimum The best condition found with constraints.

nomenclature A system of names used in arts and science.

nominal group technique A technique in which a leader poses a problem to a group and each group member then works on the problem and comes up with ideas. The ideas are discussed and then ranked.

preliminary design The step of the engineering design method where some of the more promising ideas are explored in more detail.

redundancy The use of multiple components in which each has the same capabilities.

sequential engineering An approach in which specialists work in a compartmentalized manner.

CHAPTER 6
Engineering Communications

In the curriculum for your chosen engineering discipline, you will find many "hard" courses (mathematics, science, and engineering) with just a smattering of "soft" courses (English, history, and other humanities). The hard courses emphasize computations, whereas the soft courses emphasize communication, primarily written. Given that the engineering curriculum overwhelmingly emphasizes hard courses, a student might logically conclude that communications are not important in engineering.

Nothing could be farther from the truth. The emphasis on hard courses merely reflects the necessary trade-offs that faculty must make when designing a curriculum to fit within severe time constraints. In fact, writing and oral communications are an integral part of a practicing engineer's job; some engineers report that they spend 80% of their time in these activities. Likely, these soft skills will affect the promotions of an engineer more than the hard skills, particularly if the ultimate objective is to become a manager.

A recent survey of corporations posed the following question: "What skills are lacking in recent engineering graduates?" The number-one response was that engineers lack communication skills. Because corporations are collections of individuals working toward a common goal, good communication skills are essential and highly prized.

Engineers communicate both orally and in writing. Regardless of which method, engineers use both words and graphics to present their ideas. Until now, most of your education has focused on communication via words. As a budding engineer, you also must learn to communicate graphically. Many engineering ideas are simply too complex to be described by words and can be communicated only through drawings. (Undoubtedly you have heard the expression, One picture is worth a thousand words.) Engineering graphics is a huge topic well beyond the scope of this text. There are many fine engineering graphics texts available; we trust that you will have the opportunity to study one.

In school, your goal should be to develop a set of skills that will enable you to become a successful practicing engineer. Honing your communication skills is essential to reaching that goal. Take every essay, report, and oral presentation seriously. This chapter will help you get started.

6.1 PREPARATION

Whether writing or giving an oral presentation, you must prepare by using the following three steps: topic selection, research, and organization.

6.1.1 Topic Selection

Your topic may be given, or you may be able to select it. If you are selecting your own topic, you may wish to choose something you are already familiar with or perhaps something you wish to know more about.

6.1.2 Research

There are many sources for obtaining information, as described below:

- *Technical journals* are generally devoted to a single topic (e.g., heat transfer). Authors submit their papers to journal editors who then have the papers reviewed by experts in the field. This peer review process can take a year or longer, so the results reported in technical journals are often a few years old; however, they tend to be high quality because of the review process. Technical journals are the primary means by which new information is introduced into the engineering community.

- *Books* are written by authors who are familiar with a field and wish to describe it in a consistent, coherent manner. The primary source of their information is knowledge that was first reported in technical journals, so the information in books tends to be even older than that in technical journals. The great advantage of a book is that the information is in a single source rather than in multiple articles spread over many years in a plethora of journals.

- *Conference proceedings* are a collection of papers written by authors who speak at a meeting devoted to a particular topic. Sometimes the proceedings are made available at the meeting, so the information can be extremely recent—literally, data taken a few days or weeks prior to the meeting. However, in this case, the information has not been peer reviewed, so some of the information may be of lesser quality. To overcome this problem, some conferences peer review the proceedings, but this takes time and delays publication of the information.

- *Encyclopedia articles* are very short descriptions of a particular topic. They are peer reviewed, so the information is of high quality. Like books, the information in an encyclopedia is at least a few years old.

- *Government reports* are collections of research data taken by government-sponsored researchers. The reports are required by the funding agency and are maintained at the agency. In some cases, the reports are copied onto microfilm by the National Technical Information Service, so they are more widely available. The final reports are written immediately after each project is completed, so the information is very recent; however, it usually is not peer reviewed. Generally, if the information is important to a wide audience, it will be translated into a journal article where it will be subjected to peer review.

- *Patents* describe technology that is novel, useful, and nonobvious. To be valid, a patent must fully disclose the technology so that a person "skilled in the art" can translate the patent into a working device or process. A patent protects the intellectual property of the inventor for a fixed period, typically 20 years from the time the patent was filed.

- *Popular press articles* appear in widely circulated magazines and newspapers. Usually, they are written by people with a journalism degree who have little technical background. Further, the information often must be printed quickly to meet a publication deadline, so the information may not be scrutinized. As a consequence, popular press articles dealing with technical subjects often report erroneous information. However, such articles may be useful for getting to the human side of a technical issue and may indicate how the lay public or politicians feel about controversial technical topics, such as nuclear energy, landfills, or dams. Also, they may indicate how a technical issue affects certain people, groups, or institutions.
- *Course notes* can be a good source of information; however, the information is not peer reviewed.
- *Internet sites* can be established by anyone with a computer and enough money to buy a domain name and pay monthly connection fees. The Internet is a democratic means for disseminating information. Because it lacks the scrutiny of peer review, however, erroneous information or extreme viewpoints can easily make their way into the information marketplace. Further, information on the Internet is volatile and is available only as long as the computer and its connection are maintained.

Finding information is a technical skill; in fact, libraries have experts trained to find information in their vast archives. When doing your research, freely consult these experts. But be aware that they are not experts in your field, so there are benefits to becoming more self-sufficient. The following resources will help you find information:

- *Abstracts* are brief, one-paragraph descriptions of the contents in a journal or popular press article. The abstracts are accessible through keywords or author names. There are abstracts for chemistry, biology, physics, engineering, and popular press articles (*Reader's Guide to Periodical Literature*). Abstracts are now accessible through computer searches using either CD-ROMs or the Internet.
- *Citations* or *references* are listed at the end of a publication, detailing where the information was obtained. Suppose you are a civil engineer interested in improving asphalt roads and you find a particularly good paper on asphalt chemistry by Charles Glover published in 1995. Glover's citations are an excellent way to connect to other related literature *before* 1995.
- *Citations indices* list authors and publications that have cited them. For example, knowing that Charles Glover wrote an excellent asphalt article in 1995, you can look up his name in the citations index. If another author cited his work in 1998, it is likely that this author also is writing about asphalt chemistry, so you may be interested in obtaining her paper as well. Using the citations index, it is possible to find related articles *after* 1995.
- *Library catalogs* list the holdings of the library, often by subject and author, so this is a rapid way to find relevant literature in your library.

6.1.3 Organization

When organizing your presentation or writing, the most important rule to follow is

> **Know your audience**

Again, imagine you are a civil engineer who seeks to improve our highways. If you were invited to an elementary school to speak to eight-year-olds about your work, you would certainly give a different talk than if you were invited to speak at a technical symposium on our highway infrastructure.

Once you know your audience, the next step is to determine the most important points you wish to make. Whether writing or speaking, it is usually impossible to communicate everything you know about a subject because of space or time constraints. Instead, you must carefully choose the key points and determine the most logical sequence for the ideas to flow together smoothly. An outline is helpful for achieving this goal.

When structuring your writing or speech, you may want to employ the following strategies:

- A **chronological strategy** gives a historical account of the topic. Again, imagining you are a civil engineer, you could present a history of roads by sequentially describing simple dirt paths, gravel roads, cobblestone streets, asphalt roads, and concrete multilane highways.
- A **spatial strategy** describes the component parts of an object. In the case of a road, you could describe its various features (gravel substructure, asphalt surface, drainage system).
- A **debate strategy** would describe the pros and cons of a particular approach. For example, you could describe the advantages and disadvantages of asphalt and concrete roads, with the goal of choosing which is best for a given situation.
- A **general-to-specific strategy** presents general information first, and then gives increasingly detailed information and specific examples. For example, as a civil engineer describing methods for connecting one road with another, you could first describe general considerations (e.g., number of lanes, vehicle speed, amount of traffic) and then describe specific types of connections (four-way stop signs, traffic circles, stoplights, highway cloverleafs, etc.). In some cases, a *specific-to-general* strategy is a more effective way to communicate.
- A **problem-to-solution strategy** is very effective for communicating with engineers because they are trained problem solvers. For example, a road for which you are responsible may have too many potholes. In a presentation to your boss, you could first describe this problem and then offer a variety of solutions (use a high-grade asphalt, deepen the roadbed, improve the drainage system, ban heavy trucks).
- A **motivational strategy** is often employed by sales engineers. For example, imagine you sell a high-quality asphalt that will improve road life. Your presentation could have the following components:

  1. You could get your client's *attention* by showing a picture of a pothole swallowing an automobile.
  2. You could create a *need* for your product by showing how much money your client spends fixing potholes.
  3. You could *satisfy* their need by showing how your product reduces the formation of potholes.
  4. You could help them *visualize* a better, pothole-free world.
  5. You could get them to *act* by signing a purchase contract for your asphalt product.

No matter which strategy you employ, you should have an introduction, body, and conclusion.

6.2   ORAL PRESENTATIONS

When you become an engineer, you will give oral presentations in a variety of situations—you might need to make proposals to prospective clients, explain why your company should be allowed to build a new facility in a community, explain the results of a recent analysis to your boss, or present research results at a conference. Mastering oral presentations will increase your chances of being promoted to high-visibility positions within your company.

6.2.1   Introduction

During the introduction of your oral presentation, your goal is to win your audience over. If you cannot win them over within the first few minutes, you never will. Jokes are a classic way to win the audience; if you are skilled at telling jokes, use them. However, if you are unskilled, win your audience in other ways; there is no quicker way to lose them than by telling a joke badly. Instead, you may win your audience by using anecdotes, particularly if they are personal.

During the introduction, you must connect the audience with your world. They may not have thought about your topic before, so you must grab their attention. Find an aspect of your topic that everyone can relate to. For example, everyone can relate to potholes, so this is a good way to introduce the topic of high-quality asphalt.

Commonly, the first slide of an oral presentation is the title. There is nothing wrong with this approach; however, it may not help win your audience. There is little you can do with a title but read it to the audience, which insults their intelligence. Further, many engineering presentations have titles with technical phrases that are unintelligible to most members of the audience. A more effective opening is to find an aspect of your topic that everyone can relate to so you win your audience immediately. Then, present them with enough information that they can understand every word in the title slide. Using this approach, the title slide may appear a few minutes into the presentation. This may seem awkward—but how many television shows start with the title? It is much more common to start with an opening skit that grabs the audience's attention and dissuades them from changing channels. After the skit, they show the title. This approach is very successful in television, and it can work effectively for you, too.

Often, a speaker includes an outline of the talk immediately following the title slide. Technically, this is not wrong; however, again it does little to win your audience. It is difficult to make an outline interesting, particularly if it contains technical words that few audience members understand. Instead, use "chapter" designators, which we describe next.

6.2.2   Body

The body is the heart of the presentation where you will spend the majority of your time. About 80% of the presentation is the body, with about 15% devoted to the introduction and about 5% devoted to the conclusion.

In the body, use "chapter" designators to let the audience know when you have changed topics. Suppose you were dividing your talk on road construction into the following topics: surveying, grading, roadbed preparation, and surfacing. Rather than listing these topics at the beginning of the presentation, it is more effective to have a chapter designator with the single word "Surveying" in large letters. This lets the audience know that the next few slides relate to this topic. Then, have a chapter designator with the word "Grading" in large letters to let the audience know that you have switched topics. In this way, you can step through the various topics in your presentation. An alternative approach is to indicate all the topics in a list, but highlight the particular topic being considered during the particular chapter; thus, the same list appears multiple times throughout your presentation, but each topic in the list is highlighted only once. If you organize your presentation according to a spatial strategy, a very effective chapter designator is a graphic image of the object being described. Each chapter in the presentation would begin with a portion of the graphic image highlighted. For example, you could show a cutaway view of a road indicating the soil, gravel bed, and asphalt surface. You could begin each chapter by highlighting the particular feature of the road you will discuss next.

6.2.3 Conclusion

In the conclusion, you wrap up your presentation and summarize your key points. When preparing your conclusion, think hard about what you want to be the take-home message; that is, if the audience remembers only one or two things from your presentation, what do you want those to be?

6.2.4 Visual Aids

Table 6.1 shows the results of a study in which participants were given a presentation and then later tested to determine their retention of information. This study clearly shows that combining visual aids with oral comments is the best approach.

When preparing your visual aids, employ the KISS (Keep It Simple, Stupid) principle. Each visual aid must communicate your idea rapidly; otherwise, the audience will spend too much time interpreting your visual aid and not listening to your oral comments. Obviously, you would like to make the visual aid as interesting as possible to maintain their attention. Sometimes you can accomplish this using exaggeration. Also, as much as possible, use graphic images rather than words. As the speaker, you can easily create slides containing only words; however, the audience must read the words and form a mental image all while listening to your oral comments. Most likely you will lose them.

Among the many types of visual aids are the following:

- **Word charts** convey information using short phrases, or even single words. Use bullets when the order is of no particular importance; use numbered lists when the order is

TABLE 6.1
Message retention

Presentation	Testing 3 Hours Later (%)	Testing 3 Days Later (%)
Oral only	70	10
Visual only	72	20
Oral and visual	85	65

Source: Cassagrande, D. O. and R. D. Cassagrande, *Oral Communication in Technical Professions and Businesses* (Belmont, CA: Wadsworth Publishing, 1986).

TABLE 6.2
Example of a word chart

Types of Heat Transfer

- Conduction
- Radiation
- Convection
  - —Natural convection
  - —Forced convection

TABLE 6.3
Example of a table

Typical Thermal Conductivity

Material	Thermal Conductivity (W/(m·K))
Liquid	0.1–0.6
Nonmetallic solid	0.04–1.2
Metallic solid	15–420

important. Table 6.2 shows an example of a word chart with bullets. Note that there are no complete sentences; the key is to use few words so the audience can read the chart in just a few seconds and then turn their attention to your oral comments.

- *Tables* present numerical data. Table 6.3 is an example of a table; note that both a title and units are included. Because it is difficult to visualize information in tabular form, it is better to convert the data into a chart or graph, if possible.
- *Charts* and *graphs* are images that rapidly convey numerical information. Figure 6.1 shows an example of a bar chart, a pie chart, and a graph. Note that each has a title, and units are indicated for the reported quantities.
- *Photographs* are very effective ways to communicate still images, whereas *videos* or *movies* are needed to convey moving images. All of these can add real force to a presentation.
- *Schematics* (Figure 6.2) are line drawings of objects or processes, and *illustrations* (Figure 6.3) are graphic images used to explain concepts.
- *Maps* (Figure 6.4) rapidly communicate geographic information.
- *Physical objects* can be very effective in presentations. People are interested in examining physical objects (perhaps because it brings back childhood memories of show-and-tell) and it can make the abstract become concrete. However, passing the objects can be very disruptive, so this aid works best with smaller audiences, or if done during the question-and-answer period.

Computer projections are the most popular method for presenting visual aids in technical presentations, and Microsoft PowerPoint is the most commonly used program. It is very flexible; you can change a presentation minutes before giving a talk. You can also embed eye-catching visual effects, sound effects, and even moving images in the presentation.

Here are some tips for making effective PowerPoint presentations:

- *Choose an appropriate background.* Select a background color that is pleasant to the eyes. Blue is often used. White is often used as well, but it can be glaring in a dark room. Avoid busy backgrounds that can distract from the message.
- *Select colors with care.* Use vibrant colors that stand out, but use no more than four different colors on a single slide. On a blue background, white or yellow lettering works well, whereas red or green lettering does not. Color combinationns that look good on

FIGURE 6.1

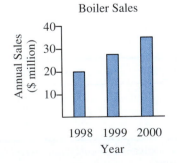

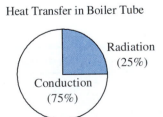

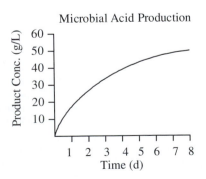

FIGURE 6.2
Example of a schematic.

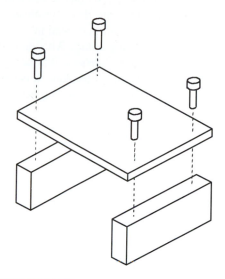

FIGURE 6.3
Example of an illustration showing how acid rain is produced from sulfur in coal.

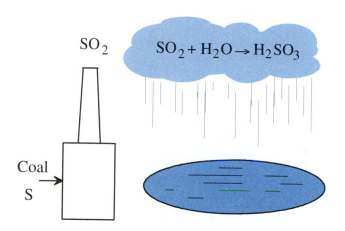

SO_2

$SO_2 + H_2O \rightarrow H_2SO_3$

Coal $\rightarrow$

S

FIGURE 6.4
Map showing locations of manufacturing facilities.

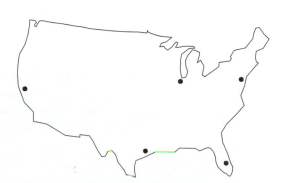

your computer screen may not look as good with the data projector. If possible, it is best to verify that your presentation works well with the actual equipment that you will use in your presentation.

- *Use graphics.* Although it is easy to create PowerPoint presentations using bulleted lists and sentences, it is very difficult for the audience to create the desired mental image from words alone. It is much better if you use compelling graphics and images. Clip art and photos are a great way to spice up your presentation, so long as they are not used excessively.
- *Simplify, simplify, simplify.* Avoid presenting too much information on a single slide; break it up into bite-sized pieces. Avoid clutter. Keep it bold and simple.
- *Be consistent.* Use the same style throughout your presentation.
- *Limit words.* Minimize the number of words. If your slide has too many words, it is not possible for the audience to both read the slide and listen to you. This causes frustration and can lose their attention. There should be no more than six to eight words per line.
- *Choose fonts carefully.* Use clear fonts at large sizes (no smaller than 24 points). Make sure your slides can be easily read from the back of the room.
- *Use titles.* Place a title on each slide. The title font should be large (no smaller than 36 points) and should be a different color than the rest of the slide.
- *Use bullets sparingly.* Use the "6 × 6" rule, meaning no more than six words per bullet and no more than six bullets per slide. It can be effective if each bullet appears one at a time rather than all at once. This allows you to control the attention of the audience. Use bullets only when appropriate, such as a list in which the order is not important. If the order is important, use a numbered list.
- *Limit capitalization.* Minimize the use of capitals; they are hard to read. Each word in your title can begin with a capital. Do not capitalize minor words (e.g., *of, in, the*) unless they are the first word in the title. The first letter of each bullet can be a capital, but the rest should be lower case (except for proper nouns).
- *Use proper spelling and grammar:* Double check your spelling and grammar. Errors show that you did not put much effort into your presentation, which sends a bad signal to your audience.
- *Refer to your slides.* Be sure to refer to your slides as you talk. Use a pointer to direct audience attention to a particular portion of your slide.
- *Avoid jargon.* Use simple words that your audience can understand.
- *Do not read slides.* Your audience is literate; they can read the slides for themselves. It is both insulting and boring to read your slides to them. The slides should have sparse text or graphics, which will not make much sense if you read them word-for-word. Your role is to embellish the slides with additional information. A possible exception is to read the slide if you have a quotation. It may be desirable to read the quotation for emphasis, and to ensure the audience is synchronized with your presentation.
- *Limit special effects.* Although you can use eye-catching visual effects—such as swooshes, dissolves, and fly-ins—it does not mean that you must use them. If overused, they can distract from your main message. If you do use a special effect to transition between slides, use the same effect throughout your presentation.

- *Be aware of differences in computer technology and setup.* If you have unusual features in your presentation, such as special symbols, typefaces, or animations, you may find that different computers respond to these features differently. If possible, it is best to use the same computer to both create and present your visuals.
- *Back up your work.* Given the complexity of computer technology, unexpected glitches can happen in the middle of your presentation that can detract from your message. To overcome this problem, it is wise to have a backup computer and backup files on a disk or flash drive.

Although computer projections are commonly used, there are other effective ways to communicate visual information, such as the following:

- *Transparencies* have the advantage of being rapidly and inexpensively reproduced by a copier or computer printer. Also, the information is available in a random-access manner, making it possible to select easily any desired transparency at any point in your presentation. Generally, color transparencies are not as vibrant as slides.
- *Slides* have stunning colors and resolution; however, they require a longer lead-time to prepare, and the information is available only in a sequential manner.
- *Blackboards and whiteboards* are effective for interacting spontaneously with the audience. However, because of the time it takes to write on them, they are not effective for rapidly conveying information. Also, there is no permanent record of the information presented this way. And, because your back is to the audience when you write, it can impede interaction with the audience.
- *Flip charts and butcher paper* are large sheets of paper that are typically mounted on an easel. They allow for spontaneous interaction with the audience and have the added advantage of maintaining a permanent record. Pre-prepared presentations can allow for rapid information transfer. Flip charts and butcher paper are favorite visual mediums in brainstorming sessions.
- *Handouts* of the presentation are often used with a small, important audience. For example, if making a sales presentation, provide copies of your presentation to the clients so they can pay attention to you rather than getting distracted by taking copious notes.

6.2.5 Speech Anxiety

In 1974, *The Bruskin Report*[1] revealed that some adults are more fearful of public speaking than financial problems, loneliness, and even death. Why would people rather die than give a speech? Most likely, the reason is that at least once in our lives, each of us has been embarrassed in front of a group. The experience may have been so humiliating that the body does not want to repeat it. Uncontrollable physiological responses to speech anxiety include sweating, shakiness, stomach distress, and an increased heart and breathing rate.

How should you respond to speech anxiety? Well, you could surrender yourself to it and become a basket case every time you must speak publicly, or you could refuse to give public speeches. Neither of these is a viable option for an aspiring engineer such as yourself; your job will require you to give oral presentations. You might try to ignore the

---

1. "What Are Americans Afraid Of?" *The Bruskin Report* (New Brunswick, N.J.: R. H. Bruskin Associates, July 1973), p. 1.

physiological symptoms, but sometimes they are simply too powerful to ignore. Instead, you should learn to harness the energy that comes from speech anxiety and direct it into your presentation.

There are a number of tricks you can use to overcome speech anxiety:

- The most powerful trick is to be well-prepared; those who are ill-prepared have good reason to be nervous.
- The first few minutes of a presentation are the most important. Recall that this is the critical time when you are trying to win the audience. It is also the time you will be most nervous, as the transition from being an anonymous audience member to becoming the center of attention puts a strain on your body. The best way to survive this initial period is to memorize the first few sentences so you can deliver them in "auto-pilot" mode. Also, take a deep breath before you utter your first words to calm your nerves.
- If you were to meet each audience member individually, you could become friends, so think of them collectively as your friends. You can reinforce this notion by picking out a few friendly faces in various parts of the room and speaking to them. Don't look at the people who are obviously bored or disinterested; they will suck the energy out of you.
- Allow yourself to make mistakes. We all trip over words or drop things. Most people in the audience will filter out your mistakes and not even notice them; however, they will notice if you respond to your mistake by becoming agitated or confused.
- The physiological manifestations of speech anxiety are based upon the "fight-or-flight" response. When your body is in a stressful situation, it is poised to do battle or to flee. Adrenaline flushes through the body, putting the nerves on edge and tightening the muscles. You can counteract these effects with endorphins that are released with intense physical exercise, such as running. By exercising one or two hours before your presentation, you will find yourself to be less nervous, which will help your presentation go well. After having many positive public speaking experiences, you will find that you get less speech anxiety each time, and the need for endorphins will eventually subside.

6.2.6 Style

According to the *Mehrabian Study*,[2] only 7% of what you communicate is verbal. The remainder is nonverbal communication, such as body language (55%) and voice strain (38%). Thus, you can be well-prepared and still give a bad presentation if your nonverbal communication is poor. Some tips on nonverbal communication follow.

- Look the audience members in the eye. It is said that the eyes are the gateway to the soul. If you will not look audience members in the eye, you will be perceived as being either a liar or a coward, neither of which is desirable. If you are nervous, you can look at their foreheads; they will never know the difference.
- As you speak, scan the audience so you are not talking to a lone friendly face. You want everyone to be involved with your presentation.
- Speak forcefully and confidently. Use your singing voice, which is supported by the diaphragm. Do not speak from the throat; doing so will make you hoarse. If you speak softly, the audience may not hear you, and may even see you as insecure.

---

2. Albert Mehrabian, *Nonverbal Communication* (Chicago, IL: Aldine Publishing Company, 1972), p. 182; Albert Mehrabian, *Silent Messages* (Belmont, CA: Wadsworth Publishing Co., 1971), pp. 43–44.

- Use a pointer to direct the audience's attention to a particular part of your visual aid. If you want the audience to stop looking at the visual aid and direct their attention to you, step away from the visual aid.
- Do not stand where you block the view to the visual aids. If you use a transparency, do not point directly to the transparency, as this usually blocks somebody's view. Instead, point directly to the screen.
- Avoid distracting habits such as jiggling change in your pocket. Do not overuse distracting phrases, such as "you know" or "ummm."
- Watch your body language. Do not be stiff, meaning you are scared, or overly relaxed, meaning you do not care. To show respect for the audience, wear nice clothing and be well groomed.
- Be enthusiastic. If you do not care about your presentation, why should the audience?

6.3 WRITING

Engineers employ technical writing, which is very different from the literary writing you learn in most English classes. Consider this passage:

> *Helen, thy beauty is to me*
> *Like those Nicèan barks of yore*
> *That gently, o'er a perfumed sea,*
> *The weary way-worn wanderer bore*
> *To his own native shore.*
>
> *Edgar Allan Poe*

To express these ideas in technical writing, we would simply say

> *He thinks Helen is beautiful.*

Although this technical writing does not elicit the emotions of Poe's passage, that is not the goal. Instead, the goals are that technical writing be

- *Accurate.* In engineering, it is essential that the information be correct.
- *Brief.* Readers of technical writing are busy and do not have time to sift through a lot of words.
- *Clear.* Be sure that your technical writing can be interpreted only one way.
- *Easy to understand.* Your goal is to express, not impress.

6.3.1 Organization

In engineering, the typical types of written communications are business letters (which are sent outside the organization), memoranda (which are sent inside the organization), proposals (which are solicitations for funding), technical reports (which are used internally), and technical papers (which are published in the open literature). Table 6.4 summarizes the content of each document. The specific order of the content and the format depend upon the organization for which you are writing.

TABLE 6.4
Contents of typical engineering documents

Business Letter	Memorandum	Proposal	Technical Report	Technical Paper
Date	To:	Title page	Title page	Title page
Recipient address	Through:	Contents	Contents	Abstract
Salutation (Dear . . .)	From:	Background	List of figures	Text
Text	Date:	Scope of work	List of tables	Introduction
Introduction	Subject:	Methods	Executive summary	Math derivations
Body	Text	Time table	Text	Methods
Conclusion	Introduction	References	Introduction	Results
Closing (Sincerely)	Body	Appendixes	Math derivations	Conclusions
Signature	Conclusion	Facilities	Methods	Acknowledgments
Stenographic reference	Enclosures	Budget	Results	Nomenclature
Enclosures		Personnel	Conclusions	References
		Letters of support	Recommendations	Appendixes
			Acknowledgments	
			Nomenclature	
			References	
			Appendixes	
1–5 pages without enclosures	1–5 pages without enclosures	1–1000+ pages	5–1000+ pages	1–40 pages

6.3.2 Structural Aids

In your written communication, be sure to employ headings and subheadings; these break the document into digestible chunks and let the reader know when you have changed topics. Paragraphs should begin with a topic sentence to inform the reader what the paragraph is about. In general, paragraphs should contain more than one sentence, unless it is presenting a particularly emphatic idea. When connecting ideas in a paragraph, be sure to use transition words, such as *but, however, in addition,* and so forth.

6.3.3 Becoming a Good Writer

Everyone has problems with writing; the problems differ only by degree. Unlike some engineering problems that you can solve by applying an algorithm that results in the correct answer every time, there is no such algorithm that guarantees good writing. Instead, you learn to write by trial and error and by reading examples of good writing. Learning to write properly requires a lifelong commitment. Slowly, over time and with practice, this skill sinks in.

Good writing requires editing; rarely does a well-written document emerge extemporaneously. The author actually must wear two hats: those of the writer and the reader. After writing a passage, you must clear your mind and read it from the viewpoint of your readers, considering their backgrounds, biases, and knowledge. (Remember: Know your audience.) Can you understand what was written, from the reader's viewpoint? This is not easy to do; after all, you just wrote it, and you know what you were *trying* to communicate. If you have the luxury of time, put the writing away for a while so you can forget what you were trying to say, and later see what you actually wrote. Alternatively, you can have someone else read your work.

Unlike natural laws, which are valid for all time and all locations, language "laws" constantly evolve. Although some grammatical rules are fairly fixed (e.g., end a sentence with a period), other rules change with time or location (e.g., "color" in the United States, "colour" in Britain). The French language has *L'Académie Française* to define proper French, but no such governing body exists for the English language. This lack is both a blessing and a curse. English freely borrows words from all over the world, allowing for many subtle shades of meaning; but then we are stuck with inconsistent spelling. Without a governing authority, English is a matter of convention. Some of the conventions are arbitrary—and some are absurd—but many allow the brain to process words rapidly and unambiguously into understanding. To impose some order on the chaotic English language, many organizations employ a style manual. Here, we describe some of the most common conventions; however, you will certainly find exceptions as you go out into the world.

6.3.4 Building Better Sentences

As you construct sentences, consider the following issues:

1. *Use parallel construction.* **When comparing related ideas, use similar sentence construction.**

   Incorrect   Scientists acquire knowledge and engineers are concerned with applying knowledge.

   Correct   Scientists are concerned with acquiring knowledge, whereas engineers are concerned with applying knowledge.

   Also employ parallel construction with lists.

   Incorrect   Civil engineers build roads, building construction, and waterworks planning.

   Correct   Civil engineers build roads, construct buildings, and plan waterworks.

2. *Avoid sentence fragments.* **Use complete sentences in your writing.**

   Incorrect   Joining a professional society, important to furthering your career.

   Correct   Joining a professional society is important to furthering your career.

3. *Use clear pronoun references.* **Be sure that the noun referenced by the pronoun is clear.**

   Incorrect   Procedure A is used for a high-concentration sample. *This* results from the benefits of advanced technology.

   Correct   Procedure A is used for a high-concentration sample. *This procedure* results from the benefits of advanced technology.

   Correct   Procedure A is used for a high-concentration sample. *This sample* results from the benefits of advanced technology.

4. *Avoid long sentences.* **Break overly long sentences into multiple short sentences.**

    Incorrect   The procedure for operating the chemical reactor starts by first opening Valve A by turning the handle counterclockwise when viewed from the top, then turning on Pump A and waiting 15 min while simultaneously watching the temperature gauge to ensure that the reactor does not overheat, in which case, open Valve B, which introduces cooling water to cool the reactor.

    Correct   The procedure for operating the chemical reactor follows: First, open Valve A by turning the handle counterclockwise when viewed from the top. Then, turn on Pump A and wait 15 min while simultaneously watching the temperature gauge. If the reactor overheats, open Valve B, which introduces cooling water to cool the reactor.

5. *Avoid short sentences.* **Combine overly short sentences into longer sentences that flow more fluidly.**

    Incorrect   The procedure for operating the chemical reactor follows: First, open Valve A. Open Valve A by turning the handle counterclockwise. The proper viewing position is from the top. Then, turn on Pump A. Wait 15 min. Simultaneously, watch the temperature gauge. If the reactor overheats, open Valve B. Opening Valve B introduces cooling water. Cooling water cools the reactor.

    Correct   (See previous item.)

    (*Note:* Occasionally using short sentences can make writing more interesting and varied.)

6. *Use active voice.* **Active sentences require fewer words and are more interesting to read.**

    Incorrect   Temperature is dependent upon the heat input.
    Correct   Temperature depends on the heat input.

    Other examples are shown in the table below:

Incorrect	Correct	Incorrect	Correct
place emphasis on	emphasize	is an indication of	indicates
is compliant with	complies with	is a representation of	represents

7. *Avoid vague words.* **Use precise words to replace general words.**

    Incorrect   The sensor read 150°F.
    Correct   The thermometer read 150°F.
    Incorrect   The community suffered through a period of economic troubles.
    Correct   For five years, the community had over 10% unemployment.

8. *Reduce prepositions.* **Overusing prepositions (*of, in, by, on, out, to, under,* etc.) makes understanding difficult.**

    Incorrect   The establishment of a panel of experts in safety was needed for investigation of accidents by miners in Pennsylvania.

Correct   A panel of safety experts was established to investigate accidents by Pennsylvania miners.

9. *Eliminate redundancies.* **Excess words take time to process and may lead to confusion.**

   Incorrect The pH value was 7.2.
   Correct   The pH was 7.2.

10. *Avoid bureaucratic language.* **The following table shows that bureaucratic phrases can be replaced by fewer words:**

Incorrect	Correct
by means of	by
in the event of	if
for the reason that	because
with regard to	about

11. *Avoid informal language.* **Using informal language is analogous to wearing jeans and a T-shirt while giving an important sales presentation.**

   Incorrect We plugged numbers into the equation.
   Correct   We substituted numbers into the equation.
   Incorrect The shaft can't rotate. (Contractions are considered casual language.)
   Correct   The shaft cannot rotate.

12. *Avoid pompous language.* **Do not use a 50-cent word when a nickel word will do the job.**

Incorrect	Correct	Incorrect	Correct
prior to	before	personnel	people
utilize	use	subsequent	next
initiate	start	terminate	end

13. *Avoid sexist language.* **In the past, if the sex of a person was indeterminate, the default sex was "he." In modern usage, "he or she" can be used, although this can lead to clunky sentences. An alternative approach is to mix "he" and "she" throughout your writing, or you can use plurals by saying "they."**

14. *Avoid dangling modifiers. Dangling modifiers* **are words or phrases that describe something that has been left out of the sentence.**

   Incorrect Determining the experiment to be a failure, the entire project was canceled.
   Correct   Determining the experiment to be a failure, the project manager canceled the project.
   Correct   Because the experiment was a failure, the entire project was canceled.

15. *Avoid split infinitives. Infinitives* **are verbs preceded by the word** *to.* **In many places, it is best to keep these two words together.**

  Undesirable   Eddie decided to quickly drive down the road.
  Preferred   Eddie decided to drive quickly down the road.

However, to provide emphasis, an infinitive may be split.

  Example   To pass the course, Edna needs to thoroughly study the class notes.

Increasingly, split infinitives are accepted unless they are awkward or ambiguous, so this is an example of a grammatical rule that is changing.

6.3.5 Punctuation

Although punctuation may seem insignificant, improper punctuation can lead to gross misunderstandings.

1. *Hyphens.* **Use hyphens to avoid ambiguity.**

  Incorrect   The specifications call for eight foot long pipes.
  Correct   The specifications call for eight foot-long pipes.
  Correct   The specifications call for eight-foot-long pipes.

Use a hyphen when spelling out fractions or numbers less than 100.

  Examples   two-thirds
                     seventy-eight

A hyphen can be used to combine nouns of equal things.

  Example   Because he does both fundamental and applied work, John is a scientist-engineer.

Use a hyphen to create compound units.

  Example   The accident rate is reported per person-mile.

Use a hyphen to create compound adjectives.

  Examples   We need a face-to-face meeting to resolve this dispute.
                     The process requires high-pressure pipe.
                     Use 5-in-diameter pipe.

  Incorrect   The pipe diameter is 5-in.
  Correct   The pipe diameter is 5 in.
  Example   This car can be powered by a six- or eight-cylinder engine.

Do not use a hyphen with adverbs ending in "ly."

  Incorrect   This is a highly-explosive process.
  Correct   This is a highly explosive process.

2. *Colons.* **Use colons to introduce a list.**

  Example   The following skills are used by engineers: analysis, creativity, and communication.

Colons can be used to introduce equations, provided a complete sentence precedes the colon.

Example The following equation results from Newton's second law:

$$F = ma$$

where

F = force
m = mass
a = acceleration

Note that a colon does not appear after the word *where*. Similarly, do not put a colon after the following words: *when, if, therefore, is, by, are, such as, especially,* and *including*.

3. *Commas.* **Use commas to separate items in a list of three or more items.**

    Example The primary tools of an engineer are a pencil, a calculator, and a computer.

    Use commas to separate nonessential or nonrestrictive clauses, that is, clauses that are parenthetical or that could be deleted without changing the meaning of the sentence.

    Example The shaft seal, which is supposed to work at high speeds, failed.

    Use commas to separate long independent clauses joined by *and, or, but,* or *nor.*

    Example In the United States we use an English measurement system, but slowly we are adopting the SI system.

    Use commas to set off introductory clauses.

    Example Before turning on the amplifier, be sure it is grounded.

    Use commas to separate multiple adjectives that could be joined by *and.*

    Example The automobile has shiny, red paint.

4. *Parentheses.* **Use to set off parenthetical lists, clarifications, acronyms, abbreviations, or asides.**

    Example Ellen's technical courses (heat transfer, fluids, and thermodynamics) are canceled.
    Example Mike suggested that we use high-pressure (schedule 80, not schedule 40) pipe.
    Example Units of measure are regulated by the National Institute of Standards and Technology (NIST).

    (*Note:* Always spell out an abbreviation upon first use.)

    Example The primary salt in seawater is sodium chloride (NaCl).
    Example The boss announced that because of his fine work, Fred will be promoted (although everyone knows it's because he married the boss's daughter).

5. *Dashes.* **Use — to emphasize parenthetical statements. (The proper symbol is —, but -- will do in a pinch.)**

    Example Open the steam valve—the one with the red handle, not the blue handle—by turning it counterclockwise.

Example   Open the steam valve--the one with the red handle, not the blue handle--by turning it counterclockwise.

6. *Semicolon.* **Do not use a comma to separate two phrases that could stand alone as independent sentences; instead, use a semicolon.**

   Incorrect   The engineering student worked hard in school, therefore he landed a good job.

   Correct   The engineering student worked hard in school; therefore, he landed a good job.

Use a semicolon to separate phrases that have commas.

   Example   The following individuals attended the meeting: Martin Fields, vice president, Ford Motor Company; Alfred Reno, chief executive officer, General Motors; and Jennifer Anderson, president, Chrysler.

7. *Apostrophe.* **To show possession for a single individual, use *'s*.**

   Example   The engineer's book was published.

To show possession for a plural noun ending in *s* or *es*, simply add the apostrophe.

   Example   When the fraternity house burned, the engineers' books were lost.

Generally, if the noun ends in an *s* sound, add *'s*.

   Example   The waitress's order

However, with some names that end in the *s* sound, simply add the apostrophe.

   Example   Gauss' law

It is best to use the possessive form only for animate creatures; however, it is sometimes employed with inanimate nouns.

   Example   The earth's orbit

8. *Quotation marks.* **Use to identify quotations.**

   Example   The astronaut's exact words were, "Houston, we have a problem."

Also, quotation marks identify a word or phrase that is used in an unconventional way.

   Incorrect   With this new computer program, click submit to file your taxes.
   Correct   With this new computer program, click "submit" to file your taxes.

In proper American usage, a comma or period appears inside the quotation.

   Incorrect   Today's dog training will concentrate on the following commands: "stay", "sit", and "down".
   Correct   Today's dog training will concentrate on the following commands: "stay," "sit," and "down."

What idiot made this rule?

6.3.6 Word Demons

"Word demons" have similar meanings or similar spellings, and are often confused with one another.

1. *Effect* (n.)*:* **result**
    Effect (v.)*:* **to cause; to bring about**
    Affect (v.)*:* **to act upon**

    Example   The effect of high-intensity noise is ear damage.
    Example   High-intensity noise effects ear damage.
    Example   High-intensity noise affects the cochlea.

2. *Compliment:* **to praise**
    Complement: **to complete or balance**

    Example   Mary's boss complimented her fine presentation.
    Example   Humanities courses complement technical courses, leading to a well-balanced education.

3. *Continuous:* **uninterrupted**
    Continual: **recurring**

    Example   Because it must operate 24 hours per day, the electric motor was designed for continuous duty.
    Example   Although it operated much of the time, the motor was plagued by continual overheating.

4. *Datum:* **single piece of information**
    Data: **multiple pieces of information**

    Example   Last night, we added another datum to our temperature log book.
    Example   These data indicate that over the past century, the average nighttime temperature has increased by 1°F.

    Note: Some style manuals treat *data* as a collective noun. In this case, the following sentence would be accepted:

    Example   The data indicates that over the past century, the average nighttime temperature has increased by 1°F.

5. *Fewer:* **used with integers**
    Less: **used with real numbers or things that cannot be counted (e.g., love)**

    Example   Because we sold fewer automobiles, our company had less income last month.

6. *Farther:* **greater physical distance**
    Further: **to a greater degree or extent**

    Example   The sun is farther away from the earth than the moon.
    Example   Further investigation showed that the engineer was unqualified to sign the construction documents.

7. *i.e.:* **abbreviation of Latin** *id est* **(that is to say)**
    e.g.: **abbreviation of Latin** *exempli gratia* **(for example)**

    Example   Engineers are better lovers; i.e., they have fewer divorces than many other professionals.

Example   A mechanical engineer must take technical courses (e.g., heat transfer, fluid flow, design).

Note: A comma always follows i.e. or e.g.

8. *Principle* **(n.): a rule, law, or code of conduct**
   Principal **(adj.): most important**
   Principal **(n.): leader, head, person in authority**

   Example   KISS (Keep It Simple, Stupid) is a basic engineering principle.
   Example   John is the principal investigator for the project.
   Example   Because of her misbehavior, Joyce was sent to see the principal.

9. *It's:* **contraction of** *it is*
   Its: **possessive of it**

   Example   It's raining outside.
   Example   Its engine overheated.

10. *There:* **that place**
    Their: **relating to them**
    They're: **contraction of** *they are*

    Example   There is the wrench.
    Example   Their wrench is on the ground.
    Example   They're going outside and taking the wrench.

11. *That:* **Restrictive: First word of a phrase that is essential for meaning.**
    Which: **Nonrestrictive: First word of a phrase that is not essential for meaning.**
    (If adding "by the way" makes sense, use *which.***)**

    Example   Of the tools that are carried on a sailboat, the screwdriver is needed most.
    Example   The screwdriver, which can also open cans of paint, is needed to tighten loose screws.

12. *Lose:* **to misplace**
    Loose: **not tight**

    Example   Do not lose that screw or we will have to find another.
    Example   That screw is loose.

13. *Since:* **from a time in the past until now**
    Because: **for the reason that**

    Example   Since the first hominid made a tool, humankind has had engineers.
    Incorrect   Since the heating element failed, the oven could no longer maintain the desired temperature.
    Correct   Because the heating element failed, the oven could no longer maintain the desired temperature.

(*Note:* This rule is not always rigidly enforced as some style manuals allow *since* to be used in place of *because*.)

14. *While:* **during the time that**
    Whereas: **although**

Correct	While studying for the exam, Fred suddenly became hungry.
Incorrect	Cargo planes are used to carry things, while passenger planes are used to carry people.
Correct	Cargo planes are used to carry things, whereas passenger planes are used to carry people.

(*Note:* This rule is not always rigidly enforced as some style manuals allow *while* to be substituted for *whereas.*)

15. *Respectively:* **in the order given**
    Respectfully: **showing respect**

Example	Experiments 1, 2, and 3 were performed on Tuesday, Thursday, and Friday, respectively.
Example	Respectfully, I disagree with your conclusion.

6.3.7 Equations and Numbers

To aid understanding, separate equations from the text line, as follows:

$$F = \frac{Q\rho}{A}$$

Notice that the numerator is written above the denominator. Although this equation could have been written $F = Q\rho/A$, this should be done only if the equation is incorporated into the text line.

Properly used italics can help clarify mathematical symbols. The following table shows when italics should, and should not, be used.

Italics	No Italics
Latin letters (*a, A, b, B,* etc.)	Greek letters (α, β, γ, etc.)
	Abbreviations (e.g., LMTD)
	Units (lb, gal, mL)
	Numbers
	Words
	Mathematical functions (e.g., sin, cos)
	Chemical formulas (e.g., NaCl)
	Parentheses ( ) and brackets [ ]

Consult Chapter 9 on the SI system for more rules on units.

For decimal numbers less than one, use a leading zero.

Incorrect	.756
Correct	0.756

In technical writing, a common dilemma is whether to report the Arabic number, or spell it out. To address this dilemma, consult the following table:

Arabic Numbers	Examples
With units of measure	The reaction takes 4 min.
	The mass is 5 g.
In mathematical or technical contexts	The voltage is 3 orders of magnitude greater.
	The velocity increased by 2 fold.
For items and sections	Use Wrench 3 to tighten the pipe.
	Section 7 shows a cross-sectional view.
For numbered objects (e.g., tables, figures, experiments)	Use the data from Experiment 3.
	Refer to Figure 4 to see the correlation.
	Table 2 shows the chemical formulas.
For all numbers in a series, even if some numbers normally would be spelled out	The tests involved 3, 8, or 15 subjects.
Ordinal numbers 10 and above	That is the 11th explosion this year.

Spell Out	Examples
Integers below 10 with no associated units	We sold eight valves today.
Ordinal numbers below 10	That is the fifth time he crashed his car.
Integers below 10, with units, if not in a mathematical or technical context	It took me seven years to write this book.
Common fractions	In today's lecture, half the students were asleep.
Numbers that begin sentences*	Fifteen valves were sold today.
Consecutive numerical expressions*	It took fifteen 2-lb weights to balance the load.

*Sometimes a sentence can be reformulated to avoid using these rules.

6.3.8 Subject/Verb Agreement

A common error in technical writing occurs when subjects and verbs do not agree.

Incorrect   The telescope and associated hardware is the major tool of astronomers.
Correct   The telescope and associated hardware are the major tools of astronomers.

The following table shows the conventions for subject/verb agreement:

Plural	Example
Compound subjects (joined by *and*)	Cattle and goats are ruminant animals.
Either/or and *neither/nor* constructions with plural nouns	Either scientists or engineers are going to fly on the space shuttle.

Singular	Example
Compound subjects that appear as a single unit	Research and development is the main activity.
Subjects modified by *each* or *every*	Each nut and screw is made of titanium.
When the subject is one of the following pronouns: *each, either, neither, one, anybody, somebody*	Neither of the animals is alive.
Collective nouns	The engineering staff is getting a raise.
Units of measure	To the beaker, 5 g of salt is added.
Either/or and *neither/nor* constructions with singular nouns	Either a scientist or an engineer is going to fly on the space shuttle.
Noun clauses that are subjects	What the schools need is donations.

6.3.9 Miscellaneous Issues

1. *Verb tense.* **In technical writing, the present tense is preferred. Use past tense only for those things that you did in the past. Consider the following examples:**

   Present   The correlation in Figure 3 shows that machining cost increases with smaller tolerances.

   Past   The flask was cleaned with chromic acid.

2. *First person.* **In formal writing, use of the first person is discouraged.**

   Informal   We studied the combustion of methane.

   Formal   The combustion of methane was studied.

   However, avoiding first person often prevents use of the active voice, so some style manuals allow use of first person, even in formal writing.

3. *Capitalization.* **Only proper nouns are capitalized.**

   Incorrect   I want to become a Mechanical Engineer.
   Correct   I want to become a mechanical engineer.

   Incorrect   To learn more about this field, visit the mechanical engineering department.
   Correct   To learn more about this field, visit the Mechanical Engineering Department.

   Incorrect   As a new hire, you are required to visit the President of the company.
   Correct   As a new hire, you are required to visit the president of the company.

   Incorrect   Our new leader is president James Garland.
   Correct   Our new leader is President James Garland.

   Naming a common noun transforms it into a proper noun. Study the following examples:

   Example   This procedure has five steps.
   Example   In this new procedure, Step 1 should be altered.

   Example   John is so efficient, he completed four experiments in a single day.
   Example   The most interesting results were reported in Experiment 3.

   Example   The reactor has 10 valves.
   Example   Cooling water is introduced by opening Valve C.

4. *Italics.* **Italics are used for foreign words employed in English sentences, scientific names of organisms, defined words, and names of long publications. (Books and journals are italicized; articles have quotation marks.)**

   Example   Well, as they say in France, *vive la différence.*
   Example   Her intestinal problems were caused by a new strain of *E. coli.*

   Example   A *heptagon* is a seven-sided figure.
   Example   Ed was elated when his article was accepted in the prestigious journal *Science.*

5. *Articles.* **The is used for a specific noun.**

   Example   The plane was overloaded when it took off, so it crashed.

A is used for an unspecified noun starting with a consonant sound.

> *Example*   If a plane is overloaded when it takes off, it will crash.

An is used for an unspecified noun starting with a vowel sound.

> *Examples*   an ulcer   a unicorn
>   an RSVP   a rebel
>   an hour   a human

6. *Figures and Tables.* **Figures and tables must be described and referenced in the text. Place the figure or table just after the first reference to it.**

> *Example*   On the next page, Figure 1 shows that costs increase exponentially with tighter tolerances.

7. *Lists.* **Start lists with simple items and proceed to the more complex items.**

> *Example*   Rosa owns the following vehicles: a bicycle, a motor scooter, and a red Mercedes with a sunroof.

8. *Spelling.* **Use a dictionary to check your spelling. Spell-check utilities can check spelling, but they are not perfect. Consider the following sentence from a student's report:**

> *Example*   Awl computer engineering students take curses in electrical circus.

9. *References.* **Each publication has its own standards for citing references. Below are some examples:**

> *Example*   Mifflin, W. B., and R. L. Jones (1978) *Engineering Design*, New York: McGraw-Hill, pp. 32–78.
>
> *Example*   Mifflin, W. B. and R. L. Jones (1978) *Journal of Engineering Design* 33, 45–64.

10. *Consistency.* **Sometimes, grammatical rules are unclear or arbitrary. In these cases, choose what you think is best and be consistent throughout your document.**

6.4 SUMMARY

Mastering engineering communications is essential. Not only will it help your career, but it could also avert disaster. Imagine the possible consequences of a poorly written operating manual for a nuclear power plant.

Before you start writing or preparing your speech, you must select a topic, conduct research, and organize your thoughts. When organizing, remember that the overwhelming consideration is, "Know your audience." Regardless of whether you are communicating orally or in writing, the communication will have an introduction, a body, and a conclusion.

For a speech, you must provide visual aids to help convey your ideas. The key is for the visual aids to communicate rapidly so the audience can focus on what you are saying;

they cannot decipher a complex slide and listen to you talk at the same time. Because graphical images are more rapidly processed by the audience, convey your thoughts graphically, if possible, rather than with written text. In oral presentations, words convey only a small part of your message; body language and voice strain convey the majority of the information.

In technical writing, the goals are accuracy, brevity, clarity, and ease of understanding. With these goals, you are fortunate to be able to communicate in English. The English language has more words than any other language in the world, allowing those who master it to convey subtle shades of meaning. As with any language, there are numerous rules and conventions that mark good writing. Those who ignore these rules send out the message that they are sloppy thinkers. If an engineering report is written poorly, the reader could logically conclude that an author who cannot master the rules of English probably cannot master technology either, a conclusion that invalidates the whole report.

Further Readings

Casagrande, D. O., and R. D. Casagrande. *Oral Communication in Technical Professions and Business.* Belmont, CA: Wadsworth, 1985.

Eisenberg, A. *Effective Technical Communication.* New York: McGraw-Hill, 1992.

————. *A Beginner's Guide to Technical Communication.* New York: McGraw-Hill, 1997.

Elliot, R. *Painless Grammar,* 2nd. ed. Hauppauge, NJ: Barron's, 2006.

Fogiel, M. *REA's Handbook of English: Grammar, Style, and Writing,* rev. ed. Piscataway, NJ: Research and Education Association, 1998.

Rozakis, L. *The Complete Idiot's Guide to Grammar and Style,* 2nd. ed. New York: Alpha Books, 2003.

Strunk, W., and E. B. White. *The Elements of Style,* 4th ed. Boston: Allyn and Bacon, 1999.

Williams, J. M. *Style: Ten Lessons in Clarity & Grace,* 8th ed. New York: Longman, 2004.

PROBLEMS

6.1 Edit the following sentences to improve the grammar and style.

(a) The CIA has spies.

(b) High temperature adversely effects metal strength.

(c) Some plant species (i.e. Drosera, Sarracenia, and Utricularia) are carnivorous.

(d) This data indicates that our reactor will blow up in 9.0 seconds.

(e) You just violated a fundamental economic principal.

(f) Please check the companies records to determine when the shipment was made.

(g) The bird sang it's song.

(h) We have the following precious metals in our safe; gold, silver, and platinum.

(i) 7 cars are in the parking lot.

(j) We just received fifty five dollar bills from the bank.

(k) We need a variable speed pump for this application.

(l) He drives a red four door car.

(m) In the past the chemical industry was based on acetylene not ethylene.

(n) The accountant, the lawyer, engineer, and physician are all professionals.

(o) Larry worked for Ford and General Motors. They gave him a raise.

(p) Open the back panel. Loosen the screw. Turn the screw counterclockwise. Use a Phillips screwdriver.

(q) During our evening watch, a shooting star was observed.

(r) The indicator showed the car was going 55 mph.

(s) From the point of view of John, this project will fail.

(t) By means of a survey, the industrial engineer improved worker efficiency.

(u) We stuck these numbers into Einstein's formula.

(v) Subsequent to that, we initiated a program to determine which personnel could be terminated in their current job function and utilized in alternate job functions.

(w) We used a jointly-developed computer program.

(x) Here, you must use a 6 in wrench.

(y) Julius has always wanted to be a Electrical Engineer.

(z) Affirming your urgency, the parts will be sent by express mail.

(aa) Diligent study being the main factor in a student's success.

(bb) In college, Mary excelled in thermodynamics, therefore, she designs jet engines as a practicing engineer.

(cc) Because of the strike, we produced less computers last month.

(dd) Of the subjects which George took in school, he found thermodynamics to be most useful.

(ee) Since the bolt was too large, it would not fit in the hole.

(ff) Petroleum engineers are responsible for finding and producing oil while chemical engineers refine oil into a variety of products.

(gg) It's mass is .789 kg.

(hh) In the circuit shown in figure 1, resistor A is a variable resistor.

(ii) Larry found the following items in his backpack: a calculus book with torn pages, a ruler, and a calculator.

6.2 A basic principle of effective writing is to eliminate unnecessary words. The following writing samples were taken from student papers. Edit them to reduce the number of words. Also, correct grammatical problems you find.

(a) A low circulation rate will reduce the load on the cyclone and decrease catalyst attrition. This can lead to much greater catalyst retention.

(b) At a temperature of 400°F, the sulfur is very viscous. This can be a problem in the first condenser, causing clogging of the tubes.

(c) After extended exposure, the catalyst accumulates coke deposits, and if left untreated, becomes deactivated. In the process, catalyst with some coke deposits is withdrawn from the reactor and sent to the continuous catalyst regenerator.

(d) The accelerators used presently are linear accelerators and cyclotrons. Linear accelerators and cyclotrons are fixed-target accelerators. These types of accelerators use a single particle accelerated at a fixed target.

(e) The coils are arranged along the walls and roof of the combustion chamber. In the combustion chamber, heat is primarily transferred by radiation.

(f) This ammonia plant is currently producing an ammonia product. It could, however, be producing more product at less cost.

(g) The more "bioballs" that are in the filter, the more effective filtration is, therefore, the biological filtration section is the largest in the filter bed.

(h) Though this device will work equally well when measuring the pressure of a gas or a liquid, matter in the liquid state will be assumed in the following discussion.

(i) There are two broad areas of Statistical Process Control; the first deals with the application of statistical techniques to the monitoring and control of industrial processes and the second deals with the inspection and acceptance of incoming material, finished product and product in progress.

(j) The measurement of the pressure in the line would be proportional to the expansion of the balloon.

(k) The third sensor discussed was a sensor involving a pipe connecting the feed line to a propeller, which, in turn, was connected to a shaft in an isobaric chamber.

(l) Hooke's law is generally applied to mechanical systems involving springs, and it states that the force exerted by the spring is directly proportional to the length it is compressed.

(m) During this time, the pressure of the hydrogen must be maintained at the proper pressure in order to avoid any possible mishaps. A sensor device located in this pipe to measure the pressure would ensure the maintaining of a suitable pressure. This pressure sensor would send an electrical signal to the process control computer.

(n) Mechanical engineers have many possible routes to take with their careers.

(o) While working they will be asked to do several kinds of jobs. They may be asked to work in design and analysis areas.

(p) To most people, aerospace engineering is thought to be working on space ships. The job of an aerospace engineer is much more than that.

(q) Computer aided engineering is affecting the entire profession because with the use of computers it is taking less time to design and correct a problem, and it is more accurate than doing everything by hand.

Glossary

chronological strategy A strategy that gives a historical account of the topic.

conference proceedings A collection of papers written by authors who speak at a meeting devoted to a particular topic.

debate strategy A strategy that describes the pros and cons of a particular approach.

general-to-specific strategy A strategy that presents general information first, and then gives increasingly detailed information and specific examples.

infinitives Verbs preceded by the word *to*.

motivational strategy A strategy that describes a problem, offers a solution, and motivates the client to act.

problem-to-solution strategy A strategy that describes the problem and offers a variety of solutions.

spatial strategy A strategy that describes the component parts of an object.

technical journal A publication devoted to a single topic.

word charts A chart that conveys information using short phrases or single words.

CHAPTER 7

Numbers

As you pursue your engineering studies, you will surely notice that numbers are everywhere. Engineers are obsessed with numbers and want to quantify everything. If an average person were to describe his new car, he would perhaps describe its color, upholstery, and stereo system. In contrast, an engineer would most likely describe its horsepower, acceleration, and weight, all of which can be quantified with numbers.

Just as a writer is concerned with accurately communicating with words, engineers are concerned with accurately communicating with numbers. Here, we describe how to use numbers properly.

7.1   NUMBER NOTATION

The U.S. standard decimal notation for numbers is

4,378.1 (U.S. standard decimal notation)

where the comma indicates three orders of magnitude and the period indicates decimals. However, in Europe, the comma replaces the period to indicate decimals and the period replaces the comma to indicate three orders of magnitude:

4.378,1 (European decimal notation)

To avoid confusion, an accepted convention is to use a space rather than a comma to indicate three orders of magnitude:

4 378.1 (Accepted convention)

Numbers written in these ways are suitable for most of the quantities we encounter in our everyday lives. However, many numbers in science and engineering are much too large or small to be recorded in decimal notation. For example, Avogadro's number (the number of molecules in a mole) would be

602,213,670,000,000,000,000,000

Because this is obviously cumbersome, **scientific notation** is generally used to represent Avogadro's number:

$$6.0221367 \times 10^{23}$$

In computers, scientific notation is often represented with a leading zero:

$$0.60221367 \times 10^{24}$$

Leading zeros are often wrongly dropped. In the popular press (magazines and newspapers), positive numbers less than one may be indicated as follows:

.593   (Popular press notation)

This convention must *never* be followed by engineers. A tiny dot separates this number from a number that is three orders of magnitude larger (593). In a world with typos and dirty copy machines, a stray dot could easily change the order of magnitude of a number. To solve this problem, engineers must ALWAYS use leading zeros for decimal numbers less than one:

0.593   (**Engineering notation**)

7.2   SIMPLE ERROR ANALYSIS

One use of numbers is for counting things. For example, if one were to ask, "How many marbles are in the following picture?"

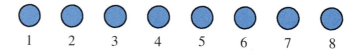

1   2   3   4   5   6   7   8

the answer is obviously the *integer* 8. Presuming there is no mistake made in the counting, the answer can be known exactly, without error.

Another use of numbers is to measure continuous properties. Assume that one were to ask the question, "What is the length of the following rod?"

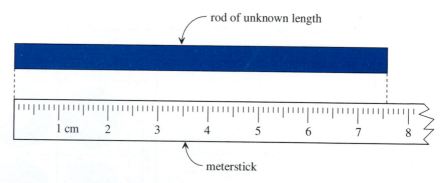

The approach to answering this question is to compare the unknown length of the rod to the known length of a meterstick. Depending on the care with which you measure the rod, you could answer using the following *real numbers:*

The rod is between 7 and 8 cm, so the length is 7.5 ± 0.5 cm.
The rod is between 7.5 and 7.6 cm, so the length is 7.55 ± 0.05 cm.
The rod is between 7.57 and 7.59 cm, so the length is 7.58 ± 0.01 cm.

If you needed to know the length precisely, you could employ more sophisticated measuring methods, such as micrometers or even lasers. No one can know the exact length of the rod because that would require an infinite number of digits. There will always be some error in the reported *real number.*

The distinction between *integers* and *real numbers* is very important in computers. Computers can represent integers exactly, provided the number does not exceed the limits of the machine. However, when computers manipulate real numbers (in computer jargon, the computer is performing "floating-point operations"), there are inherent errors, because an infinite number of digits are required to represent real numbers exactly.

Whenever measurements are made, important distinctions arise such as **accuracy** versus **precision, systematic errors** versus **random errors,** and **uncertainty** versus **error.** The differences between these concepts are a source of confusion.

- *Accuracy* is the extent to which the reported value approaches the "true" value and is free from error (Figure 7.1).
- *Precision* is the extent to which the measurement may be repeated and the same answer obtained (Figure 7.1).
- *Random errors* result from many sources, such as random noise in electronic circuits and the inability to reproducibly read instruments. For example, it is very difficult to read the meterstick the same way every time. Even though you may close one eye and try to read the number scale from a perpendicular position, you may report a slightly different measurement each time you read the scale.
- *Systematic errors* result from a measurement method that is inherently wrong. Consider an engineer who finds an old rusty scale made of steel and iron. He has a service technician clean it, calibrate it, and certify that it correctly weighs a set of standard weights. The engineer uses the scale for many months and comes to trust its measurements. One day, while measuring the weight of a very powerful magnet, he notices the scale is giving an unusually high reading. He then realizes there is a systematic error; the magnet

FIGURE 7.1
Accuracy and precision of bullets hitting a target.

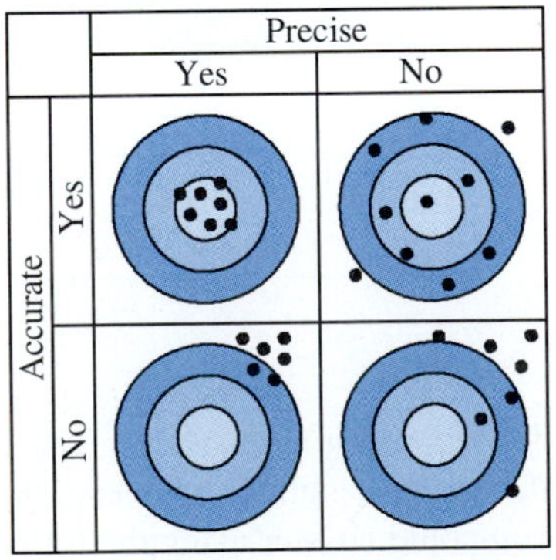

is attracted to the iron and steel components of the scale, which causes a high reading. To eliminate this systematic error, he must use a scale constructed from nonmagnetic materials such as plastic or stainless steel.

- *Uncertainty* results from random errors and describes the lack of precision. The uncertainty in the rod measurement may be expressed on a fractional or percentage basis:

$$\text{Fractional uncertainty} = \frac{\text{uncertainty}}{\text{best value}} = \frac{0.01 \text{ cm}}{7.58 \text{ cm}} = 0.0013 \qquad (7\text{-}1)$$

$$\text{Percentage uncertainty} = \frac{\text{uncertainty}}{\text{best value}} \times 100\% = \frac{0.01 \text{ cm}}{7.58 \text{ cm}} \times 100 = 0.13\% \qquad (7\text{-}2)$$

Note that the rod length was precise to about 1 part in 1000, which is typical of most measurements we make.

- *Error* may be defined as the difference between the reported value and the true value:

$$\text{Error} = \text{reported value} - \text{true value} \qquad (7\text{-}3)$$

Error results from systematic errors and describes the lack of accuracy. To determine the true value, it is necessary to correct the systematic error. The error may be reported as the fractional error or the percentage error:

$$\text{Fractional error} = \frac{\text{error}}{\text{true value}} \qquad (7\text{-}4)$$

$$\text{Percentage error} = \frac{\text{error}}{\text{true value}} \times 100\% \qquad (7\text{-}5)$$

EXAMPLE 7.1

Problem Statement: An Arctic researcher measures the length of a rod at $-60°C$. She records the length to be 7.58 cm. Her assistant notes a systematic error, in that the aluminum meterstick was calibrated at room temperature ($20°C$), not $-60°C$. Because the meterstick shortened at the cold temperature, the true length of the rod is actually shorter than 7.58 cm. The assistant goes online and learns that aluminum shortens by 23.6×10^{-6} cm/cm for each °C that it is cooled.

She calculates that the rod shortened by 0.0143 cm, so the actual length of the rod is 7.57 cm. What is the fractional error and the percentage error?

Solution:

$$\text{Fractional error} = \frac{0.0143 \text{ cm}}{7.57 \text{ cm}} = 0.0089$$

$$\text{Percentage error} = \frac{0.0143 \text{ cm}}{7.57 \text{ cm}} \times 100\% = 0.89\%$$

Trouble with Hubble

Hubble is a space-based telescope that is designed to observe distant galaxies without interference from earth's atmosphere. Soon after it went into operation, astronomers were disappointed by its blurry images. They had great expectations for this telescope; the 2.4-m (94.5-inch) diameter mirror was **precisely** smooth to within 9.7 nm (0.00000038 inch). To put this smoothness into perspective, if the Hubble mirror were the diameter of the Gulf of Mexico, ripples on its surface would be only 0.5 cm (0.2 inch) high.

The main problem with Hubble originated from a **systematic error.** During the manufacture of the mirror, the null corrector (an instrument that measures mirror geometry) was improperly positioned. The mirror was supposed to be ground to a perfect hyperbolic shape. However, because it was not **accurately** made, the shape deviates slightly from a perfect hyperbola; the edges are about 2 μm (0.00008 inch) too low, making it too flat. If it were a perfect hyperbola, all the light would focus onto the light collector (see figure), but because of the defect, the light did not focus properly, thus greatly reducing the light collected and giving star images a "halo." Thus, the Hubble mirror was **precisely inaccurate.** Fortunately, because it was precisely made, the distortion was later corrected by modifying the light collector.

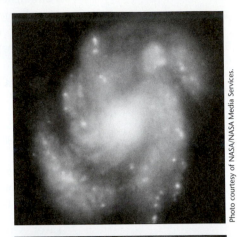

Photo courtesy of NASA/NASA Media Services.

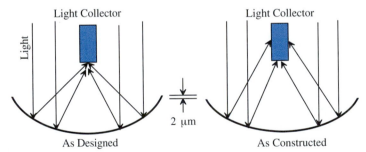

Hubble images of the M100 galactic nucleus. Top image before correction, bottom image after correction.

Suppose we wished to express the length of the rod in meters, rather than centimeters. Recalling that one meter is exactly 100 centimeters (by definition), we can convert the measured length into meters:

$$7.58 \text{ cm} \times \frac{1 \text{ meter}}{100 \text{ cm}} = 0.0758 \text{ m}$$

Note that the original measurement had two decimal places, whereas the new reported measurement has four decimal places. Clearly, the new measurement is no more accurate than the original measurement, so the number of decimal places is completely irrelevant as a measure of accuracy. Rather than describing accuracy according to the number of decimal places (as is often erroneously done), the concept of **significant figures** is required. Both the above numbers have three significant figures, so they both convey the same degree of accuracy.

7.3 SIGNIFICANT FIGURES

The issue of *significant figures* is central to how we use numbers and how much we believe them. A common question is, How many significant figures should I use when reporting a number? The answer to the question depends on how well you know the number. The following table should help you decide.

If You Know the Number to:	Then Report This Many Significant Figures:
1 part per 10	1
1 part per 100	2
1 part per 1000	3
1 part per 10,000	4
1 part per 100,000	5
1 part per 1,000,000	6
etc.	etc.

The uncertainty in the rod length measurement was about 1 part per 1000, so three significant figures are appropriate. The accepted convention is that the last reported digit has some error, whereas the first digits are known exactly.

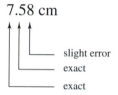

The following rule of thumb should help you decide how many significant figures to report:

> Many engineering measurements are accurate to 1 part in 1000, so three significant figures are appropriate. For estimates, only one or two significant figures should be reported. If a measurement is made using an extremely accurate instrument, then four or more significant figures are warranted.

To determine the number of significant figures in a number, use the following rules:

- A *significant figure* is an accurate digit, although the last digit is accepted to have some error.
- The number of significant figures does not include the zeros required to place the decimal point.

The latter rule causes some confusion. By looking at a number, it is not always possible to determine if the zeros are truly significant, or whether they are simply locating the decimal point. Consider these examples:

Number	Number of Significant Figures
0.00342	3
342	3
340	2 or 3

The first two numbers clearly have three significant figures, whereas the last number is ambiguous. It is impossible to determine whether the last zero is truly significant, or whether it is needed to locate the decimal place. (*Note:* Some conventions use a decimal point to show significant figures. As an example, 340 is understood to have two significant figures, whereas 340. is understood to have three. Unfortunately, this convention is not universally followed. Also, it can be ambiguous. For example, the number 3400. would be interpreted to have four significant figures, but how many figures are significant in 3400?) To avoid ambiguity, scientific notation should be used.

Number	Number of Significant Figures
3.42×10^{-3}	3
3.42×10^{2}	3
3.40×10^{2}	3
3.4×10^{2}	2

Two important points must be made about significant figures:

1. Exact definitions have an infinite number of significant figures. For example, one inch is **defined** to be **exactly** 2.54 centimeters,

$$1.00000000000^{+} \text{ inch} \equiv 2.5400000000000^{+} \text{ centimeters}$$

where the superscript "+" indicates there are an infinite number of zeros. Generally, the zeros would not be written; it is understood that there are an infinite number of them.

2. Numbers resulting from exact mathematical relationships have an infinite number of significant figures. For example, the area A of a circle can be calculated from the radius r as

$$A = \pi r^2$$

The radius is exactly half the diameter; therefore,

$$A = \pi \left(\frac{D}{2} \right)^2 = \frac{\pi}{4} D^2$$

In this formula, the 4 is equivalent to 4.0000000^{+} and the 2 is equivalent to 2.0000000^{+}.

Numbers are *rounded* when there are more digits than are appropriate. For example, calculators generally display many more digits than are significant, so the final answer must be rounded. The following rules determine how to properly round:

- Round up if the number following the cut is between 5 and 9.
- Leave alone if the number following the cut is between 0 and 4.

This is better understood by considering the following examples:

Calculator Display	Desired Number of Significant Figures	Reported Number
5,937,458	3	5,940,000
0.23946	3	0.239
0.23956	3	0.240

It should be emphasized that the rounding should be performed only when the final answer is reported. Do not round during the intermediate calculations. Some students diligently round at each calculation step, thus introducing so much error that the final answer is wrong. This is called **rounding error.**

Rounding error is often encountered in computer calculations, because computers represent real numbers with a finite number of digits. The adverse effects of rounding error can be reduced by declaring variables as "double precision," meaning that twice the normal number of digits are used to represent real numbers. Some programming languages even support "variable precision," which allows the user to represent real numbers with as many digits as they require.

When real numbers are used in calculations, the uncertainty in the final answer is dictated by the most uncertain real number. There are sophisticated techniques for determining how the uncertainties (or errors) propagate in the calculation. Here, we present some simple rules to help you determine how many significant figures can be reported in the final answer.

Significant Figures: Multiplication/Division.
Consider the following example for $A \times B = C$:

		Calculator Display	Reported Value
Lower bound	$4.9 \times 10.623 =$	52.0527	
"Best" value	$5.0 \times 10.624 =$	53.1200	53 ± 1
Upper bound	$5.1 \times 10.625 =$	54.1875	

From this example, it is clear that the two-digit number limits the final reported number to two significant figures. The appropriate procedure for multiplying/dividing numbers follows:

1. Indicate the number of significant figures for each number.
2. Calculate the answer.
3. Round the answer to have the same number of significant figures as the least precise number.

$$(2) \quad (5) \qquad\qquad (2)$$

$$5.0 \times 10.624 = 53.120 \Rightarrow 53$$

Significant Figures: Addition/Subtraction.
Consider the following calculation for $A + B = C$:

		Calculator Display	Reported Value
Lower bound	4.9 + 14.696 =	19.596	
"Best" value	5.0 + 14.697 =	19.697	19.7 ± 0.1
Upper bound	5.1 + 14.698 =	19.798	

From this example, it is clear that the two-digit number limits the accuracy of the reported value. But notice that the reported value actually has *three* significant figures. The appropriate procedure for adding/subtracting numbers follows:

1. Align the decimal points.
2. Mark the last significant figure of each number with an arrow.
3. Calculate the answer.
4. The arrow farthest to the left dictates the last significant figure of the answer.

$$
\begin{array}{r}
5.0 \downarrow \\
+14.697 \\
\hline
19.697 \rightarrow 19.7
\end{array}
$$

For mixed calculations, use the algebraic order of operations:

1. Parentheses
2. Exponents
3. Multiplication and division
4. Addition and subtraction

During each step in the calculation, keep track of the significant figures.

EXAMPLE 7.2

Problem Statement: Perform the following calculation and report the appropriate number of significant figures:

$9.5987 \times (2.3 + 89.257)$

Solution:

$9.5987 \times (91.557)$

$9.5987 \times (91.6)$

879.24092

879

7.4 SUMMARY

Numbers are indicated according to a variety of conventions. In the United States, a period indicates the decimal, and commas mark three orders of magnitude. This convention is switched in Europe. Very large and very small numbers are often written in scientific notation. Decimal numbers between -1.0 and $+1.0$ are indicated with a leading zero.

Numbers may be classified as integers (used for counting) or reals (used to measure continuous properties). Integers have no error if the counting is done correctly. In contrast, real numbers do have errors, because an infinite number of significant figures are required to represent them exactly. An accurate real number means it is very close to the true value, whereas a precise real number means that the same number is obtained after repeated measurements. Random errors make a measured number less precise. Even though a measurement may be precise, systematic errors may cause it to be inaccurate. Error (i.e., the difference between the measured value and the true value) results from systematic errors. Uncertainty results from random errors.

The better a number is known, the more significant figures are reported. Numbers resulting from exact definitions, or exact mathematical relationships, have an infinite number of significant figures. When performing mathematical operations on real numbers, it is important to report the final answer with an appropriate number of significant figures.

Further Reading

Eide, A. R., R. D. Jenison, L. L. Northup, and S. K. Mikelson. *Engineering Fundamentals and Problem Solving,* 5th ed. New York: McGraw-Hill, 2008.

PROBLEMS

7.1 How many significant figures are reported in each of the following numbers?
(a) 385.35
(b) 0.385×10^3
(c) 40 001
(d) 40 000
(e) 4.00×10^4
(f) 0.400×10^4
(g) 389,592
(h) 0.0000053
(i) 345
(j) 3.45

7.2 Round the following numbers to three significant figures.
(a) 356,309
(b) 0.05738949
(c) 0.05999999
(d) 583,689
(e) 3 556
(f) 0.004555

(g) 400,001
(h) 730 999

7.3 Perform the following computations and report the correct number of significant figures. Ensure that the reported answer is rounded properly.
(a) 39.4×3.4
(b) $39.4 \div 3.4$
(c) $39.4 + 3.4$
(d) $39.4 - 3.4$
(e) $(0.0134)(5.58 \times 10^2)$
(f) $(248,287 \text{ in}^2)(\text{ft}^2/144 \text{ in}^2)$
(g) $(452 \text{ cm})(\text{m}/100 \text{ cm})$
(h) $(34.7 - 49.0456)/7$
(i) $(0.00034)(48,579) - 345.984$
(j) $x^2 + 3.4x + 3.982$ where $x = 9.4$
(k) $4.0568 \times 10^{-3} - 0.492 \times 10^{-2}$
(l) $8.9245 \times 10^4/6.832 \times 10^{-5}$

7.4 An experienced engineer is designing a bridge to cross a river. He needs to know the distance between two fixed points on opposite banks of the river. He assigns a young engineer the task of measuring the distance between the two points. The young engineer purchases a very long tape measure and nails the end to one fixed point. He then gets in a boat and rows to the other bank while taking the tape measure with him. When he finds the other fixed point on this bank, he stretches the tape measure as tightly as he can and measures a distance of 163 meters. When the experienced engineer hears how the measurement was made, he laughs and says that the measurement is completely inaccurate because the tape measure sags in the middle because of gravity. No matter how tightly it is stretched, the sag will still introduce too much error. The experienced engineer recommends using a laser range finder, which bounces a laser pulse off a mirror placed on the opposite bank and measures the time it takes to return. Because gravity has a negligible effect on light, this method can measure the straight-line distance between the two points. The laser range finder reports the distance as 138 meters. What is the fractional error and percentage error of the original measurement?

7.5 A laser range finder consists of a laser and a light receiver. It operates by bouncing a laser pulse off a mirror and measuring the "time of flight," the time it takes for the pulse to leave the laser, bounce off the mirror, and return to the receiver. The speed of light in a vacuum is exactly 299,792,458 m/s, so the accuracy of the distance measurement depends strictly on the accuracy of the time measurement. Using a laser range finder, an engineer measures the time of flight to be 3.45 ± 0.03 μs. [*Note:* A microsecond (μs) is 10^{-6} s.]

(a) Derive a formula for the distance between the mirror and the laser/receiver, and calculate the distance using the measured time of flight.

(b) What is the fractional uncertainty and percentage uncertainty in the time of flight?

(c) What is the fractional uncertainty and percentage uncertainty in the distance measurement?

(d) What is the uncertainty (in meters) in the distance between the mirror and the laser/receiver?

(e) The speed of light in air is 0.02925% slower than in a vacuum. If a correction were applied to account for this, is it correcting for random error or systematic error?

(f) Do you think it is necessary to correct the reported distance measurement because the speed of light in air is slightly slower than in a vacuum?

7.6 A European engineer works for an American automobile manufacturer. She measures the diameter of an automobile drive shaft using a micrometer, an accurate measuring device. The microme-ter reads 2.0573. When she records the reading, she instinctively writes "2.0573 cm." An American engineer is checking the European's work and determines there is a mistake; the microme-ter measures inches, not centimeters. What is the fractional error and percentage error of the diameter recorded by the European engineer?

7.7 Avogadro's number is measured by determining the number of atoms in exactly 0.012 kg of carbon 12. As you might imagine, it is very difficult to count individual atoms, so there is necessarily some error in the reported value. Other sources of error are con-taminants in the carbon 12 and the difficulty of measuring exactly 0.012 kg. The reported value is 6.0221367×10^{23}. The "true" value probably lies between 6.0221295×10^{23} and 6.0221439×10^{23}. (There is only one chance in twenty that the true value would be outside this range.) What is the fractional uncertainty and percent-age uncertainty in the reported value of Avogadro's number?

7.8 Write a computer program that allows the user to input the "true" value and the "reported" value. The program calculates and reports the fractional error and the percentage error. Use the data presented in Problem 7.4.

7.9 Write a computer program that multiplies two real numbers A and B and calculates the answer C. The two inputs A and B may have an arbitrary number of significant figures (up to 8). The reported answer C must have the correct number of significant fig-ures as determined by the rules for multiplication/division. The program may ask the user how many significant figures are in A and B, but it must calculate the number of significant figures in C.

7.10 Write a computer program that accomplishes the same task as Problem 7.9, but the program may not ask the user how many significant figures are in A and B. The program must determine the number of significant figures from the user input. For example, if the user says A is 89.43, the program must determine that there are four significant figures.

7.11 Write a computer program that adds two real numbers A and B and calculates the answer C. The two inputs A and B may have an arbitrary number of significant figures (up to 8). The reported answer C must have the correct number of significant figures as determined by the rules for addition/subtraction. The program may ask the user how many significant figures are in A and B, but it must calculate the number of significant figures in C.

7.12 Write a computer program that accomplishes the same task as Problem 7.11, but the program may not ask the user how many sig-nificant figures are in A and B. The program must determine the number of significant figures from the user input. For example, if the user says A is 63, the program must determine that there are two significant figures.

Glossary

accuracy The extent to which the reported value approaches the "true" value and is free from error.

engineering notation The use of leading zeros for decimal numbers less than 1.

error The difference between the reported value and the true value.

precision The extent to which the measurement may be repeated and the same answer obtained.

random error An error that does not result from a measurement method that is inherently wrong.

rounding error Rounding in intermediate calculations, which results in an incorrect final answer.

scientific notation Numbers expressed in terms of a decimal number between 1 and 10 multiplied by a power of 10.

significant figure An accurate digit, excluding the zero required to place the decimal point.

systematic error An error that results from a measurement method that is inherently wrong.

uncertainty The result from random errors and describes the lack of precision.

CHAPTER 8

Tables and Graphs

Many professions, such as law, rely almost exclusively on the written and oral word. Although engineers also must write and speak well, this alone is insufficient to convey complex engineering information. For this, graphical or visual communication is required. In just a few seconds, a well-prepared graph can accurately communicate information that would require many pages of written text. In addition, a graph can provide readers with insight they can obtain through no other means. Graphs are prepared from tabulated data, so understanding tables goes hand-in-hand with understanding graphs.

Presenting graphical information in a coherent, visually appealing manner is one of the engineering arts. Mastering it will help your career tremendously. Although personal style and taste affect graphical presentations, we will present the rules and guidelines that are almost universally agreed upon.

8.1   DEPENDENT AND INDEPENDENT VARIABLES

We generally understand nature to work in a cause/effect manner. When an engineer studies a system, he often classifies some variables as *independent* (i.e., cause) and other variables as *dependent* (i.e., effect). For example, let's assume an engineer is studying an automobile (the system) and is interested in the factors that affect its speed. The **dependent variable** in this case is speed s. Some **independent variables** that affect the speed are the rate of fuel entering the engine $f;$ tire pressure $p;$ air temperature $T;$ air pressure $P;$ road grade $r;$ car mass $m;$ frontal area $A;$ and drag coefficient C_d. This can be mathematically stated as

$$s = s(f, p, T, P, r, m, A, C_d) \tag{8-1}$$

which says the automobile speed depends on variables such as fuel rate, tire pressure, and so forth. Obviously, because an automobile is a complex system, a simple algebraic formula will not describe the functional relationship between the dependent variable and independent variables. Sophisticated computer modeling or experimentation will be required.

8.2   TABLES

A table is a convenient way to list dependent and independent variables. The independent variable(s) are usually listed in the left column(s) and the dependent variable(s) are usually listed in right column(s). The values in a given row correspond to each other.

For example, let's assume the engineer actually did an experiment using a car. He measured the car speed at different fuel rates and road grades. All the other parameters were kept constant. A properly constructed table is shown in Table 8.1. The independent variables (road grade and fuel rate) are listed in the left two columns, and the dependent variable (speed) is listed in the right column. The data were taken by driving the car on three roads, one flat (0% grade) and the other two on hills (5 and 10% grades). On each road, the gas pedal was adjusted to give various fuel rates from 1 to 5 gal/h, as measured by an electronic flow meter installed in the fuel line. The speed was measured by an electronic instrument that reports the speed to an accuracy of 0.1 mi/h. The objective is to produce a table that is self-contained, so the reader does not have to read accompanying text to fully understand the experiment. Therefore, other relevant parameters are specified in the table. For example, the car model is identified, which specifies the frontal area, drag coefficient, and mass. The tire pressure, and air temperature and pressure are also indicated.

Notice that the numbers presented in the table have an appropriate number of significant figures. The instruments were capable of reporting three significant figures, which is typical of most engineering instruments. The decimal points are vertically aligned, making

TABLE 8.1
Properly constructed table

Effect of Fuel Rate and Road Grade on Car Speed — Title

Car Model = XLR
Tire Pressure = 30 psig
Air Temperature = 70°F
Air Pressure = 0.985 atm

— Additional Information

Road Grade (%)	Fuel Rate (gal/h)	Speed (mi/h)
0.00	1.00	38.2
	2.00	64.3
	3.00	81.0
	4.00	93.4
	5.00	99.2
5.00	1.00	32.2
	2.00	54.5
	3.00	68.4
	4.00	78.1
	5.00	84.8
10.00	1.00	25.8
	2.00	43.6
	3.00	54.7
	4.00	62.5
	5.00	67.8

Column Heads / Units / Numerical Values

Each axis is graduated with *tick marks*. It is preferred that the tick marks appear outside the graph field so that they do not interfere with the data.

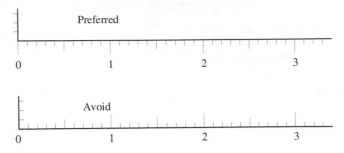

The numbers on the axes should be spaced so they can be easily read. Compare the following two axes. Obviously the first one is easier to read.

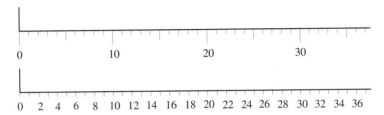

The smallest graduations on the scale are selected to follow the *1, 2, 5 rule,* meaning that if the number were written in scientific notation, the mantissa would be a 1, 2, or 5. The following axes show some examples of acceptable and nonacceptable graduations:

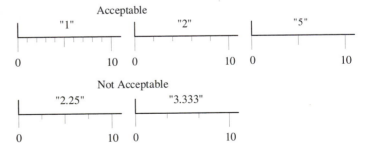

The allowable exceptions to the 1, 2, 5 rule include units of time (days, weeks, years, etc.) because these are not decimal numbers.

Problems can result if the numbers on the axis are extremely large or small, because they will crowd.

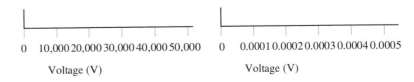

This problem can be solved in a number of ways. The best method is to use the SI system multipliers (see Chapter 9). In the SI system, "k" means "1000 ×" and "m" means "0.001 ×." Therefore, these axes can be cleaned up as follows:

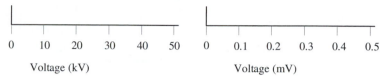

Voltage (kV)   Voltage (mV)

It is **extremely** important to observe the case of the units and multipliers. For example, if an uppercase "M" is used instead of the lowercase "m," the meaning is completely changed. "M" means "1,000,000 ×," so this would change the meaning by 9 orders of magnitude!

The use of the SI multipliers is convenient for solving the problem of numbers that crowd together. However, many engineering units do not have multipliers; and, by tradition, some SI units do not have multipliers (e.g., °C). There are two conventions to solving this problem. These conventions are not universally followed or understood; therefore, it is incumbent upon the reader to know the meaning from the context.

The first convention uses numbers or words as multipliers, rather than SI symbols.

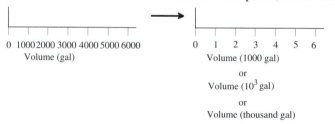

The second convention is to take the actual value and multiply it by an appropriate number so that the reported value has fewer digits.

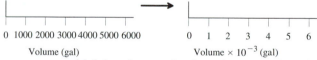

Note that here we are multiplying the actual volume by ten to the **minus** three so that the number reported on the axis is three orders of magnitude smaller. The natural tendency is for the reader to take the reported value and multiply by 10^{-3} giving a "milligallon" (if such a unit existed). The reader would be wrong by 6 orders of magnitude! Because these two conventions are so similar, it is easy for the reader (and author) to be confused. It is best to check the number from the context and determine if it makes sense. If an error were made, it would be huge; errors of this magnitude can usually be caught using good thinking.

Another problem often encountered in graphs is how to plot numbers that span many orders of magnitude. For example, what if you wanted to have the following numbers on the abscissa (x-axis):

2, 23, 467, 3876, and 48,967

This problem is solved by using a logarithmic scale rather than a linear scale.

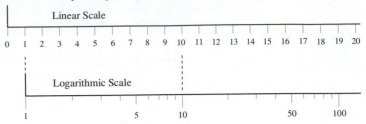

This particular logarithmic scale has 2 orders of magnitude, so it is called a *two-cycle* log scale. If it had 3 orders of magnitude, it would be called a *three-cycle* log scale. Notice that the log scale has no zero; it occurs at minus infinity on the linear scale.

Data points are plotted with symbols. The following symbols are commonly used:

○ ● □ ■ △ ▲ ▽ ▼ ◇ ◆ Yes

Complex symbols such as these should be avoided:

⊕ ⊗ ⊙ ⊞ ⊠ ✛ ⬖ ✕ ⊡ ⊡ No

The reader must look at each complex symbol very carefully to distinguish among them. This is difficult because graphs are often printed in a small format. Also, many photocopiers blur the symbols. An open circle with a dot in it could easily become a closed circle when blurred, thus leading to confusion. The size of the symbols is also very important. They must be large enough for the reader to easily distinguish them, but not so large that they run into each other.

A different symbol is used for each data set. For example, the car speed data were taken using three different road grades, so it would be appropriate to use three different symbols, one for each road grade.

The data points are often connected together with lines. Different line styles are commonly encountered:

Although the line style may also be used to differentiate data sets, the preferred procedure is to differentiate data sets with different data point symbols and connect the symbols with solid lines of uniform width. It should be noted that lines must not penetrate into open symbols, because they could easily be mistaken for closed symbols.

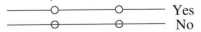

The meanings of the symbols or lines must be identified on the graph. This is generally done in one of three ways: (1) in the figure title, (2) in a legend, or (3) adjacent to the lines. The third method is preferred, because the reader can instantly identify the meaning of the symbols/lines without having to make a connection between the symbols listed in the legend or figure title and those presented in the graph.

Data may be categorized as: observed, empirical, or theoretical. *Observed* data are often simply presented without an attempt to smooth them or correlate them with a mathematical model [Figure 8.2(a)]. *Empirical* data are presented with a smooth line, which may be determined by a mathematical model, or perhaps it is just the author's best judgment of where the data points would have fallen had there been no error in the experiment [Figure 8.2(b)]. *Theoretical* data are generated by mathematical models [Figure 8.2(c)]. Note that data points are shown with both observed and empirical data but **not** with theoretical data. No data points are indicated with theoretical data because the calculated points are completely arbitrary and of no interest to the reader.

FIGURE 8.2

Data for a chemical reaction: (a) observed, (b) empirical, and (c) theoretical.

(a)

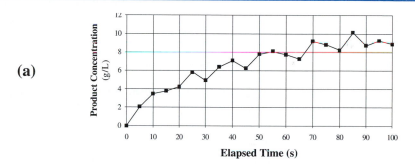

(b)

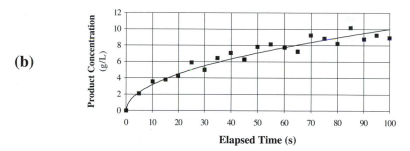

(c)

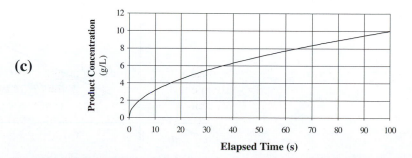

8.4   LINEAR EQUATIONS

Figure 8.3 shows that the two distinct points (x_1, y_1) and (x_2, y_2) establish a straight line. The point (x, y) is an arbitrary point on the line.

The **slope,** m, of this line is defined as "rise over run," or

$$m = \frac{y_2 - y_1}{x_2 - x_1} \tag{8-2}$$

Because all the quantities in this equation are known, the slope can be calculated. (*Note:* This equation is valid only for $x_2 \neq x_1$ to avoid division by zero. If $x_2 = x_1$, then the line is vertical, with the equation $x = x_1 = x_2$.)

The equation for the slope may also be written by using the arbitrary point (x, y):

$$m = \frac{y - y_1}{x - x_1} \tag{8-3}$$

Both sides of this equation may be multiplied by $(x - x_1)$, yielding

$$y - y_1 = m(x - x_1) = mx - mx_1 \tag{8-4}$$

$$y = mx + y_1 - mx_1 \tag{8-5}$$

If the constant b is defined as $(y_1 - mx_1)$, this equation becomes

$$y = mx + b \tag{8-6}$$

The constant b is interpreted as the *y-***intercept** because $x = 0$ when $y = b$. The *x-intercept, a,* is where $y = 0$. From Equation 8-6, it is easy to show that

$$a = -\frac{b}{m} \tag{8-7}$$

FIGURE 8.3
The straight line established by points (x_1, y_1) and (x_2, y_2).

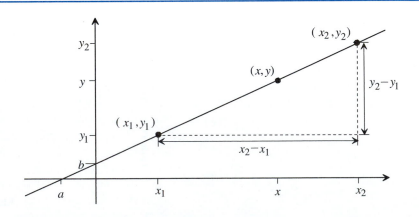

8.5 POWER EQUATIONS

A **power equation** has the form

$$y = kx^m \tag{8-8}$$

If a logarithm is taken of both sides, then the power equation becomes linear:

$$\log y = \log(kx^m) = \log(x^m k)$$

$$\log y = \log x^m + \log k$$

$$\log y = m \log x + \log k \tag{8-9}$$

Thus a plot of log y versus log x gives a straight line with a slope m and y-intercept log k, which is analogous to b in Equation 8-6. One can derive this linear equation using any desired base (2, e, or 10).

If the exponent m is positive, then the power equation plots as a *parabola*. Figure 8.4(a) shows a plot of

$$y = 2x^{0.5} \tag{8-10}$$

using linear axes. This equation becomes linear when logarithms are taken of both sides:

$$\log y = 0.5 \log x + \log 2 \tag{8-11}$$

The following table shows some selected values of x and y, along with the corresponding logarithms.

x	y	log x	log y
1	2.000	0.000	0.301
2	2.828	0.301	0.452
3	3.464	0.477	0.540
5	4.472	0.699	0.651
10	6.325	1.000	0.801
25	10.000	1.398	1.000

When plotting these data, we have a choice. We may plot y versus x directly on a **log-log graph** [Figure 8.4(b)], or we may plot log y versus log x on a **rectilinear graph** [Figure 8.4(c)]. The advantage of a log-log graph is that x and y may be read directly from the axes. Also, it eliminates the need to calculate the logarithms; in effect, the logarithmic axis is doing the calculation for you. **Unfortunately, the slope of the log-log graph is not meaningful.** (See this for yourself. Determine the slope at two places on the line, and you will see that they differ.) The advantage of the rectilinear graph is that the slope is meaningful. A rectilinear graph is particularly useful for plotting experimental data where the exponent m will be determined from the measured slope and the constant k will be determined from the measured y-intercept.

Astute readers will notice that in Figure 8.4, "data" points are plotted from the **calculated** numbers in the table. Properly, as stated in Section 8.3, data points should not

(a)

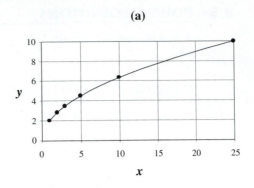

(b)

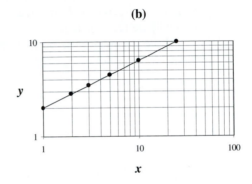

(c)

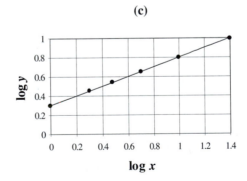

FIGURE 8.4

Plots of the parabolic equation $y = 2x^{0.5}$.

be indicated because calculated numbers are not really data. Here, for teaching purposes, we violate the rule to show you how the numbers in the table plot on the figure.

If the exponent m is negative, then the power equation plots as a *hyperbola*. Figure 8.5(a) shows a plot of

$$y = 10x^{-0.8} \tag{8-12}$$

(a)

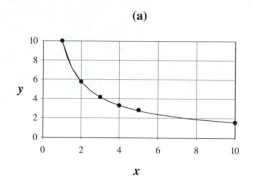

(b)

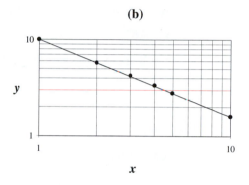

(c)

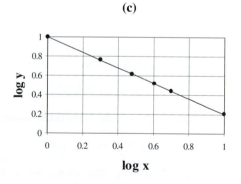

FIGURE 8.5

Plots of the hyperbolic equation $y = 10x^{-0.8}$.

using linear axes. This equation becomes linear when logarithms are taken of both sides:

$$\log y = -0.8 \log x + \log 10 \tag{8-13}$$

The following table shows some selected values of x and y, and the corresponding logarithms.

x	y	log x	log y
1	10.000	0.000	1.000
2	5.743	0.301	0.759
3	4.152	0.477	0.618
4	3.299	0.602	0.518
5	2.759	0.699	0.441
10	1.585	1.000	0.200

Figure 8.5(b) shows y versus x on a log-log graph, and Figure 8.5(c) shows $\log y$ versus $\log x$ on a rectilinear graph.

8.6   EXPONENTIAL EQUATIONS

An **exponential equation** has the form

$$y = kB^{mx} \tag{8-14}$$

where B is the desired base (e.g., 2, e, or 10). Assuming base 10 is used, this equation becomes

$$y = k10^{mx} \tag{8-15}$$

Logarithms (base 10) are taken of both sides to give a linear equation:

$$\log y = \log(k10^{mx}) = \log(10^{mx}k)$$

$$\log y = \log 10^{mx} + \log k$$

$$\log y = mx + \log k \tag{8-16}$$

Thus, a plot of $\log y$ versus x gives a straight line with slope m and intercept $\log k$, which is analogous to b in Equation 8-6.

Figure 8.6(a) shows a plot of

$$y = 6 \times 10^{-0.5x} \tag{8-17}$$

using linear axes. It becomes linear when logarithms are taken of both sides:

$$\log y = -0.5x + \log 6 \tag{8-18}$$

The following table shows some selected values of x and y, and the corresponding value of $\log y$.

x	y	log y
0	6.000	0.778
1	1.897	0.278
2	0.600	−0.222
3	0.190	−0.722
4	0.060	−1.222
5	0.019	−1.722

(a)

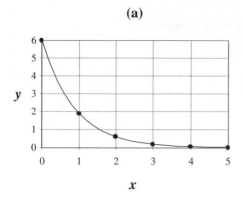

(b)

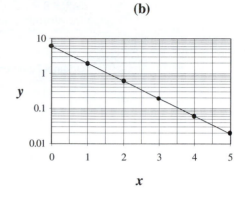

(c)

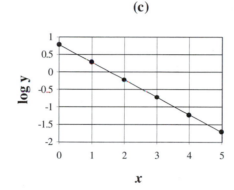

FIGURE 8.6
Plots of the exponential
equation $y = 6 \times 10^{-0.5x}$.

Figure 8.6(b) shows y versus x on a **semilog graph,** and Figure 8.6(c) shows log y versus x on a rectilinear graph. The semilog graph has the advantages that y can be read directly and that there is no need to calculate log y because the axis does the calculation. However, the slope is meaningless (as with log-log graphs). If experimental data are being plotted to determine the slope and intercept, it is necessary to plot the data on a rectilinear graph.

8.7 TRANSFORMING NONLINEAR EQUATIONS INTO LINEAR EQUATIONS

To an engineer, there is something marvelous about a straight line. One can always argue about a curve; does it fit this equation or that? But if the data plot is a straight line, there is no question; a straight line is a straight line. Therefore, engineers often manipulate nonlinear equations into a linear form, as you have already seen with the power and exponential equations.

Table 8.2 shows some examples of nonlinear equations that may be transformed into linear equations through appropriate manipulations.

TABLE 8.2
Transforming nonlinear equations into linear equations

Original Equation	Redefined Equation				Graph
	y	m	x	b	
$c = a \sin q + d$	c	a	$\sin q$	d	
$c = aq^2 + f/g$	c	a	q^2	f/g	
$\dfrac{1}{c} = \dfrac{q^2 - 3}{a} + f/g$	$\dfrac{1}{c}$	$\dfrac{1}{a}$	$q^2 - 3$	f/g	
$c = \dfrac{1}{aq^2 + f}$	$\dfrac{1}{c}$	a	q^2	f	

EXAMPLE 8.1

Problem Statement: If a liquid and vapor coexist in the same vessel and come to equilibrium (i.e., the temperature and pressure are the same everywhere in the vessel and do not change with time), the pressure in the vessel is called the **vapor pressure** (Figure 8.7).
   A nonlinear equation predicts the vapor pressure,

$$P = 10^{A\,-\,B/T} \tag{8-19}$$

where P is the vapor pressure, T is the absolute temperature, and A and B are constants. Figure 8.8(a) shows vapor pressure data for water (steam) as a function of temperature. Manipulate this equation so that it plots linearly.

Solution: We can transform this equation into a linear equation by taking the logarithm of both sides, writing

$$\log P = A - \frac{B}{T} = A - B\left(\frac{1}{T}\right) \tag{8-20}$$

Thus, a plot of $\log P$ versus $1/T$ will be linear [Figure 7.8(b) and (c)] with slope $-B$ and intercept A.

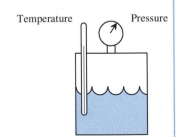

Temperature   Pressure

FIGURE 8.7
Vapor pressure exerted
by a liquid.

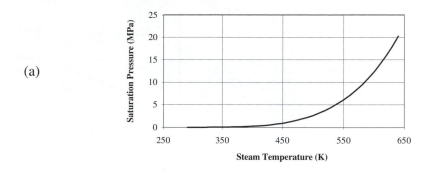

(a)

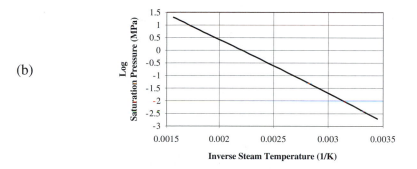

(b)

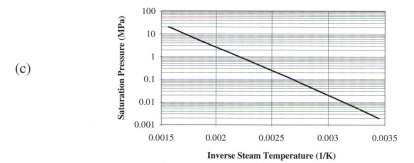

(c)

FIGURE 8.8
Vapor pressure of water (steam).

8.8   INTERPOLATION AND EXTRAPOLATION

Interpolation is extending between the data points and **extrapolation** is extending beyond the data points (Figure 8.9). The smooth curve drawn between the data points is actually an interpolation, because there are no data between the points. Provided there are a sufficiently large number of closely spaced data points, interpolation is safe. Extrapolation, on the other hand, can be quite risky, particularly if the extrapolation extends far beyond the data.

Linear interpolation approximates a curve with a straight line (Figure 8.10). A straight line passes through the points (x_1, y_1) and (x_2, y_2) which are on the curve. Provided

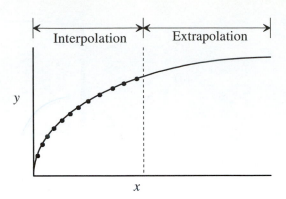

FIGURE 8.9
Interpolation and extrapolation.

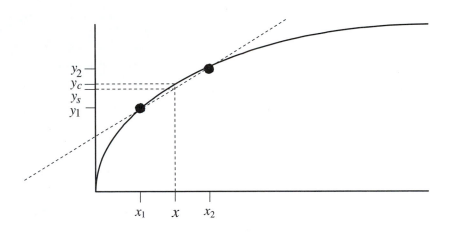

FIGURE 8.10
Linear interpolation.

these points are close to each other and the curve is continuous, the straight line approximates the curve well. We can be more explicit and say that for an arbitrary value of x that lies between x_1 and x_2, the corresponding value of y that lies on the curve (y_c) is very close to the corresponding value of y that lies on the straight line (y_s). The values for y that lie on the straight line are easily calculated from the equation for a straight line,

$$y_s = mx + b \qquad (8\text{-}21)$$

The slope m is easily determined from the two points that lie on the curve, by

$$m = \frac{y_2 - y_1}{x_2 - x_1} \qquad (8\text{-}22)$$

The y-intercept is determined by substituting this expression for the slope and solving for b at the known point (x_1, y_1).

$$y_1 = \left(\frac{y_2 - y_1}{x_2 - x_1}\right)x_1 + b \tag{8-23}$$

$$b = y_1 - \left(\frac{y_2 - y_1}{x_2 - x_1}\right)x_1 \tag{8-24}$$

Because we now have equations for the slope and intercept in terms of known quantities, we can substitute these into the equation for a line and derive a formula for y_s in terms of an arbitrary x.

$$y_s = \left(\frac{y_2 - y_1}{x_2 - x_1}\right)x + \left(y_1 - \frac{y_2 - y_1}{x_2 - x_1}x_1\right) \tag{8-25}$$

$$y_s = y_1 + \left(\frac{y_2 - y_1}{x_2 - x_1}\right)(x - x_1) \tag{8-26}$$

This final expression allows calculation of y_s, which is approximately equal to the desired value y_c.

Linear interpolation may also be performed on tabulated data. Suppose there is a table that has data in the range of interest, but the specific value you are seeking is not listed. You know the independent variable x and are seeking the dependent variable y.

Independent Variable	Dependent Variable
x_1 fractional difference x x_2	y_1 same fractional difference y y_2
x_3	y_3
⋮	⋮

Using linear interpolation, we can say the fractional difference between the dependent variables is the same as the fractional difference between the independent variables. Mathematically, this is stated as

$$\text{Fractional difference} = \frac{x - x_1}{x_2 - x_1} = \frac{y - y_1}{y_2 - y_1} \tag{8-27}$$

This may be solved explicitly for y:

$$y = y_1 + \left(\frac{y_2 - y_1}{x_2 - x_1}\right)(x - x_1) \tag{8-28}$$

Note that this equation is identical to the one developed from the graphical approach to linear interpolation (Equation 8-26).

EXAMPLE 8.2

Problem Statement: The *steam tables* list properties of water (steam) at different temperatures and pressures. Because water is one of the most common materials on earth, these tables are widely used by a variety of engineers. Among the properties listed in the steam tables is the vapor pressure of water at various temperatures. Suppose you wish to know the vapor pressure of water at 114.7°F. Although this temperature is not explicitly listed in the table, estimate it by linear interpolation.

Temperature (°F)	Pressure (psia)
112	1.350
114	1.429
116	1.512
118	1.600
120	1.692

Solution: $P = 1.429 + \dfrac{1.512 - 1.429}{116 - 114}(114.7 - 114) = 1.458 \text{ psia}$

8.9 LINEAR REGRESSION

In mathematics, we are normally given a formula from which we calculate numbers. If we reverse this (i.e., determine the formula from the numbers), the process is called **regression,** meaning "going backward." If the formula we seek is the equation of a straight line, then the process is called **linear regression.** (*Nonlinear regression* seeks an equation other than that for a straight line. This is an advanced topic beyond the scope of this book. A number of commercially available computer programs are able to perform nonlinear regression.)

A linear regression problem may be stated as follows: Given a data set, what are the slope and y-intercept that best describe these data? Two approaches are generally taken, the *method of selected points* and **least-squares linear regression.**

One starts the *method of selected points* by plotting the data (Figure 8.11) and then manually drawing a line that best describes the data as judged by the person analyzing the data. In principal, the person analyzing the data is attempting to draw a line that is close to **all** the data points, not just a few. Two arbitrary points at opposite ends of the line are selected. (*Note:* These points are **not** necessarily data points; they are merely two convenient points that fall on the line.) These two selected points (x_1, y_1) and (x_2, y_2) are substituted into Equations 8-22 and 8-24 so that the slope and y-intercept may be calculated.

EXAMPLE 8.3

Problem Statement: A group of engineering students who study together decide they must use their time more efficiently. Because they have many classes, they must allocate their study time to each class in an optimal manner. They decide that an equation that predicts their exam grade on the basis of number of hours studied will help them allocate their time more efficiently. Based upon the following data set, what linear equation correlates their performance? Use the method of selected points.

Student	Hours Studied	Grade
1	5	63
2	10	91
3	2	41
4	8	75
5	6	69
6	12	95
7	0	32
8	4	50
9	8	80

Solution:

The following graph shows the plotted data.

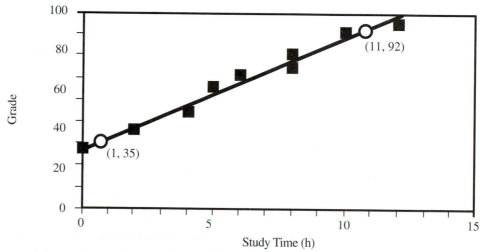

The line was drawn by eye and passes through the following points: (1, 35) and (11, 92). The slope and intercept are calculated as follows:

$$m = \frac{y_2 - y_1}{x_2 - x_1} = \frac{92 - 35}{11 - 1} = 5.7$$

$$b = y_1 - mx_1 = 35 - 5.7(1) = 29.3$$

The problem with the method of selected points is that it relies on the judgment of the person analyzing the data. If 100 people were to analyze the data, there likely would be 100 different slopes and 100 different y-intercepts. The next method is not affected by personal bias; all 100 people would produce the same "best line" and the same slope and y-intercept.

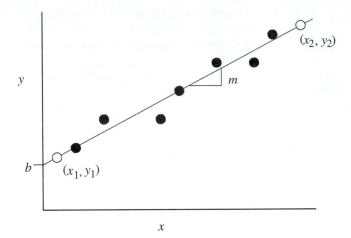

FIGURE 8.11

Method of selected points.

Least-squares linear regression uses a rigorous mathematical procedure to find a line that is close to all the data points (Figure 8.12). The difference between the actual data point y_i and the point predicted by the straight line y_s is the residual d_i:

$$d_i = y_i - y_s \tag{8-29}$$

$$d_i = y_i - (mx_i + b) \tag{8-30}$$

The residual will assume both positive and negative values, depending on whether the data point is above or below the line; however, the square of the residual d_i^2 is always a positive number. The "best" straight line that describes the data would have the smallest possible d_i^2. This is rigorously stated as "find m and b such that the sum d_i^2 is a minimum,"

$$\text{Sum} = \sum_{i=1}^{n} d_i^2 = \sum_{i=1}^{n} [y_i - (mx_i + b)]^2 \tag{8-31}$$

where n is the number of data points. Performing this minimization is the subject of differential calculus and is beyond the scope of this book. Here the results are simply presented for two cases:

1. Best line: $y = mx + b$

$$m = \frac{n(\Sigma x_i y_i) - (\Sigma x_i)(\Sigma y_i)}{n(\Sigma x_i^2) - (\Sigma x_i)^2} \tag{8-32}$$

$$b = \frac{\Sigma y_i - m(\Sigma x_i)}{n} \tag{8-33}$$

2. Best line through the origin: $y = mx$

$$m = \frac{\Sigma x_i y_i}{\Sigma x_i^2} \tag{8-34}$$

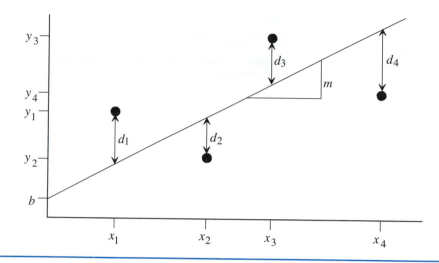

FIGURE 8.12

Least-squares linear regression.

The *correlation coefficient r* is used to determine how well the data fit a straight line. It is defined as

$$r \equiv \pm \sqrt{1 - \frac{\Sigma(y_i - y_s)^2}{\Sigma(y_i - \bar{y})^2}} \tag{8-35}$$

where $\bar{y}$ is the mean value of y, defined as

$$\bar{y} \equiv \frac{\Sigma y_i}{n} \tag{8-36}$$

In Equation 8-35, the positive sign is used for a positively sloped line, and the negative sign is used for a negatively sloped line. If all the data lie precisely on the line, then $\Sigma(y_i - y_s)^2 = 0$ and $r = 1$ (positively sloped line) or $r = -1$ (negatively sloped line). If the data are scattered randomly and do not fit a straight line, then $\Sigma(y_i - y_s)^2 = \Sigma(y_i - \bar{y})^2$ and $r = 0$.

Although Equation 8-35 is useful for understanding the meaning of r, it is inconvenient to use. The following equation is an alternative version of r that is more convenient:

$$r \equiv \frac{n(\Sigma x_i y_i) - (\Sigma x_i)(\Sigma y_i)}{\sqrt{n(\Sigma x_i^2) - (\Sigma x_i)^2} \sqrt{n(\Sigma y_i^2) - (\Sigma y_i)^2}} \tag{8-37}$$

EXAMPLE 8.4

Problem Statement: The engineering study group described in Example 8.3 decides that they would like a more accurate equation that describes the relationship between grades and study time. They prepare the table shown in Table 8.3 on the basis of their performance on the last exam. Using least-squares linear regression, what equation correlates their performance? What is the correlation coefficient?

TABLE 8.3
Effect of studying on grades

Student	Hours Studied (x_i)	Grade (y_i)	(y_i^2)	(x_iy_i)	(x_i^2)
1	5	63	3969	315	25
2	10	91	8281	910	100
3	2	41	1681	82	4
4	8	75	5625	600	64
5	6	69	4761	414	36
6	12	95	9025	1140	144
7	0	32	1024	0	0
8	4	50	2500	200	16
9	8	80	6400	640	64
$n = 9$	$\sum x_i = 55$	$\sum y_i = 596$	$\sum y_i^2 = 43{,}266$	$\sum x_iy_i = 4301$	$\sum x_i^2 = 453$

Solution:

$$m = \frac{9(4301) - (55)(596)}{9(453) - (55)^2} = 5.64 \tag{Equation 8-32}$$

$$b = \frac{(596) - 5.64(55)}{9} = 31.8 \tag{Equation 8-33}$$

$$r = \frac{9(4301) - (55)(596)}{\sqrt{9(453) - (55)^2}\,\sqrt{9(43{,}266) - (596)^2}} = 0.98878 \tag{Equation 8-37}$$

As you can see, this calculation is rather tedious. Fortunately, many calculators with statistical packages can perform linear regression with just a few keystrokes.

EXAMPLE 8.5

Problem Statement: An engineer determines the spring constant k of a spring by applying a force F, causing it to compress (shorten) by a displacement d. The engineer correlates the data with the equation

$$F = kd \tag{8-38}$$

The engineer will therefore determine the best line that goes through the origin. She collected the data in Table 8.4. What is the best value for the spring constant? What is the correlation coefficient?

Solution:

$$k = m = \frac{2150.2}{204.0} = 10.5 \text{ kN/mm} \tag{Equation 8-34}$$

$$r = \frac{8(2150.2) - (36.00)(378.8)}{\sqrt{8(204.0) - (36.00)^2}\,\sqrt{8(22{,}663.82) - (378.8)^2}} = 0.999996 \tag{Equation 8-37}$$

TABLE 8.4
Spring displacement due to applied force

Measurement	Displacement† (mm) (x_i)	Force† (kN) (y_i)	(y_i^2)	(x_iy_i)	(y_i^2)
1	1.00	10.3	106.09	10.3	1.00
2	2.00	20.8	432.64	41.6	4.00
3	3.00	31.3	979.69	93.9	9.00
4	4.00	42.1	1772.41	168.4	16.0
5	5.00	52.7	2777.29	263.5	25.0
6	6.00	63.2	3994.24	379.2	36.0
7	7.00	73.9	5461.21	517.3	49.0
8	8.00	84.5	7140.25	676.0	64.0
$n = 9$	$\sum x_i = 36.00$	$\sum y_i = 378.8$	$\sum y_i^2 = 22633.82$	$\sum x_iy_i = 2150.2$	$\sum x_i^2 = 204.0$

† Experimentally, force is the independent variable and displacement is the dependent variable. However, Equation 8-38 treats displacement as the independent variable and force as the dependent variable; therefore, displacement is shown in the left column and force is shown in the right column.

8.10   SUMMARY

Tables list corresponding independent and dependent variables on the same row. A properly constructed table has a descriptive title and columns that are identified with both heads and units. Although tables are very useful for putting data into a computer or calculator, they are difficult to comprehend and interpret; for this, graphs are required.

A properly constructed graph has a descriptive title and each axis is identified with a label and units. Engineers typically construct three types of graphs: rectilinear, semilog, and log-log. A linear equation plots as a straight line on rectilinear graphs, an exponential equation plots as a straight line on semilog graphs, and a power equation plots as a straight line on log-log graphs. When regressing data (i.e., determining an equation from data), it is best to use rectilinear graphs. If the data fit a linear equation, a plot of y versus x is straight; if the data fit an exponential equation, a plot of log y versus x is straight; and if the data fit a power equation, a plot of log y versus log x is straight. (*Note:* Alternatively, the natural logarithms may be used.) Many other equations will plot linearly on rectilinear graphs by properly manipulating the variables. For example, log P versus $1/T$ will plot linearly if P is the absolute vapor pressure and T is the absolute temperature of a liquid in equilibrium with vapor.

Interpolation is the process of extending between data points and extrapolation is the process of extending beyond the data points. Although extrapolation is fairly risky, interpolation is generally safe. Most commonly, linear interpolation is used.

Linear regression is a process of finding the best straight line that fits the data. The method of selected points may be used to "eyeball" the data and find the best straight line. More rigorous is the method of least-squares linear regression.

Further Readings

Eide, A. R., R. D. Jenison, L. L. Northup, and S. K. Mikelson. *Engineering Fundamentals and Problem Solving*, 5th ed. New York: McGraw-Hill, 2008.

Felder, R. M., and R. W. Rousseau. *Elementary Principles of Chemical Processes*. 3rd ed. New York: Wiley, 1999.

PROBLEMS

8.1 A *resistor* is a device that converts electrical energy into heat by passing electric current through a poor electrical conductor (e.g., carbon) rather than a good electrical conductor (e.g., copper). Electrons (i.e., electric current) will flow through the resistor when a voltage, such as that from a battery, is applied (Figure 8.13).

The current I through a resistor is proportional to the amount of applied voltage V:

$$I = \left(\frac{1}{R}\right) V$$

where the inverse resistance, $1/R$, is the proportionality constant.

A single battery cell has a 1.5-volt output. By placing the battery cells in series (by stacking them end to end), it is possible to obtain voltages that are multiples of 1.5 volts. The voltages and currents in Table 8.5 were measured for a resistor with an unknown resistance.

Plot the data using rectilinear axes with current on the y-axis and voltage on the x-axis. The slope of this line is the inverse resistance, $1/R$. Determine the slope by two methods:

(a) Method of selected points.

(b) Least-squares linear regression.

What is the value of the resistance? (*Note:* The unit of electrical resistance is the ohm (Ω), which is identical to volt/ampere.) What is the correlation coefficient? You may use a spreadsheet if you wish. (*Hint:* Current flows *only* when voltage is applied.)

TABLE 8.5
Voltage drop through a resistor

Applied Voltage (volt)	Current (ampere)
1.5	0.11
3.0	0.26
4.5	0.35
6.0	0.50
7.5	0.61
9.0	0.68
10.5	0.81
12.0	0.92
13.5	1.02

8.2 As a car gets heavier, fuel consumption increases because of greater friction losses in wheels, and because more kinetic energy is converted into heat each time the car brakes to a stop. Since the energy crisis in the 1970s, automakers started downsizing cars, making them lighter by decreasing their size and substituting plastic and aluminum for steel. Of course, other factors affect fuel consumption, such as aerodynamic shape and engine design.

The data in Table 8.6 were obtained from a random sampling of automobiles. Using rectilinear axes, plot fuel consumption versus car weight. Determine the slope and y-intercept of a straight line that correlates the data, using:

(a) Method of selected points.

(b) Least-squares linear regression.

What is the correlation coefficient? You may use a spreadsheet if you wish.

8.3 The decay of a radioactive element is described by the equation

$$A = A_0 e^{-kt}$$

where A is the amount at time t, A_0 is the amount at time zero, and k is the decay constant. The data in Table 8.7 were taken for the highly radioactive element balonium-245. Plot the following:

FIGURE 8.13

Electric circuit in which current flows through a resistor.

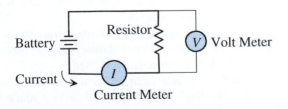

TABLE 8.6
Effect of automobile mass on fuel consumption

Automobile Mass (lb_m)	Fuel Consumption (miles per gallon)
2534	24.3
3023	15.9
2294	30.7
3797	12.5
2876	20.4
2382	35.8
3498	22.8
2475	40.3
2103	45.5

TABLE 8.7
Decay of balonium-245

Time (day)	Balonium-245 (grams)
0.00	45.3
0.05	30.4
0.10	20.9
0.15	14.1
0.20	9.4
0.25	6.1
0.30	4.1
0.35	3.0
0.40	2.0

(a) Balonium versus time using rectilinear axes.
(b) Napierian logarithm of balonium versus time using rectilinear axes (calculate correlation coefficient).
(c) Balonium versus time on semilog paper.
Determine the decay constant, k. (*Note:* The units of k are days^{-1}.) You may use a spreadsheet if you wish.

The *half-life* is defined as the time it takes for half of the balonium-245 to decay. Alternatively, we could say it is the time at which half the balonium-245 still remains, or $A/A_0 = 0.5$. What is the half-life of balonium-245?

8.4 John, a skeptical engineering student, reads in a physics book that "any object, regardless of its mass, has the same acceleration a_0 when it falls under the influence of earth's gravity in the absence of other forces. Provided the object starts from rest, the position x

as a function of time t is given by the equation $x = \frac{1}{2}t^2$." The engineer doubts this statement because he knows that if you drop a rock and a leaf from the same height, the rock will hit the ground first. To test the claim made in the physics book, he decides to do an experiment. He enlists the help of six friends. After much discussion, they decide that the most practical way to make the measurements is to drop a baseball and a cannonball down the center shaft of the stairwell in their dormitory. Each friend will stand on a different floor while holding a stopwatch. When John releases the ball from the top floor, he yells "NOW," and each friend starts his stopwatch. Each friend agrees to stop his watch when the ball passes by his feet. (They were going to stop the watch when the ball passed at eye level, but because they are all different heights, they realized this would introduce systematic error into their experiment.) With a tape measure, they determine that the floors are separated by twelve feet. After some practice, they take the data in Table 8.8.

TABLE 8.8
Distance traveled by dropping a baseball and cannonball

Distance (ft)	Baseball time (s)	Cannonball time (s)
12	0.85	0.87
24	1.24	1.22
36	1.43	1.55
48	1.73	1.75
60	1.93	1.93
72	2.20	2.15

Using their data, prove to yourself that the physics equation is correct by doing the following:
(a) Plot distance versus time using rectilinear axes.
(b) Plot distance versus time on log-log paper.
(c) Plot $\log_{10}$ distance versus $\log_{10}$ time using rectilinear axes (calculate the correlation coefficient for the baseball data and cannonball data).
(d) Plot ln distance versus ln time using rectilinear axes.
(e) Determine the acceleration and the exponent for time using the method of selected points.
(f) Determine the acceleration and the exponent for time using least-squares linear regression.
You may use a spreadsheet to make your plots, if you wish.

On the basis of the results of the experiment, do you agree with the physics book that all objects have the same acceleration under the pull of earth's gravity regardless of their mass? If John had performed the experiment with a rock and a leaf, what results would you expect and why?

8.5 After performing the experiment described in Problem 8.4, John believes the equation given in his physics book is correct. He would like to revise his estimate of the acceleration by interpreting the data with the time exponent exactly equal to 2. He does not know how to do this, so he comes to you for help. How would you plot the data? Using this new approach, what is your revised estimate of the acceleration due to gravity?

8.6 In the following equations, y is the dependent variable, x is the independent variable, and a and b are constants. Indicate what type of graph you would use to obtain a straight line (i.e., rectilinear, semilog, log-log). Also, indicate what will be plotted on the ordinate and the abscissa.

(a) $y = ax$
(b) $1/y = ax + b$
(c) $y = (ax + b)^{-1}$
(d) $y = (ax + b)^{-2}$
(e) $y = (ax + b)^2 + 5$
(f) $y = a10^{bx}$
(g) $y = ae^{bx}$
(h) $y = a2^{bx}$
(i) $y^2 = 3 + a10^{b(x-4)}$
(j) $y = \left[3 + \left(ae^{\frac{b}{x-4}}\right)\right]^{-2}$
(k) $y = ax^b$
(l) $y = [a(x - 5)^b]^{-1}$
(m) $\sin y = ax^b$
(n) $y = a(\cos x)^{-b}$
(o) $y = [ax^b]^{-1/2}$
(p) $y = 4 - \left(\dfrac{1}{ax^b}\right)$

8.7 The following variables plot linearly on a log-log graph. Develop an equation that relates the variables.
(a) y versus x (Answer: ax^b where a and b are constants.)
(b) $1/y$ versus $1/x$
(c) $1/(y - 3)$ versus x
(d) $\sin y$ versus $1/x$
(e) $1/y^2$ versus x

8.8 The following variables plot linearly on a semilog graph. Develop an equation that relates the variables.
(a) y versus x (Answer: $y = ae^{bx}$ or $y = a10^{bx}$ where a and b are constants.)
(b) $1/y$ versus $1/x$
(c) $1/(y - 3)$ versus x
(d) $\sin y$ versus $1/x$
(e) $1/y^2$ versus x

8.9 Determine the equation $y = f(x)$ for each of the following cases. Report the equation in its simplest form. All of the plots are straight lines. All coordinates are indicated with the abscissa first and the ordinate second, i.e., (x, y).

(a) y versus x on a rectilinear graph passes through $(5, 7)$ and $(2, 3)$.
(b) $\ln y$ versus $\ln x$ on a rectilinear graph passes through $(5, 7)$ and $(2, 3)$.
(c) $\log y$ versus $\log x$ on a rectilinear graph passes through $(5, 7)$ and $(2, 3)$.
(d) y versus x on a log-log graph passes through $(5, 7)$ and $(2, 3)$.
(e) $\ln y$ versus x on a rectilinear graph passes through $(2, 3)$ and $(4, 6)$.
(f) $\log y$ versus x on a rectilinear graph passes through $(2, 3)$ and $(4, 6)$.
(g) y versus x on a semilog graph passes through $(2, 3)$ and $(4, 6)$.
(h) y^2 versus x on a rectilinear graph passes through $(3, 2)$ and $(6, 4)$.
(i) $\sqrt{y}$ versus x on a rectilinear graph passes through $(3, 2)$ and $(6, 4)$.
(j) y^2 versus x on a log-log graph passes through $(3, 2)$ and $(6, 4)$.
(k) $\sqrt{y}$ versus x on a log-log graph passes through $(3, 2)$ and $(6, 4)$.
(l) y^2 versus x on a semilog graph passes through $(3, 2)$ and $(6, 4)$.
(m) $\sqrt{y}$ versus x on a semilog graph passes thorugh $(3, 2)$ and $(6, 4)$.
(n) $(y - 2)^2$ versus x on a rectilinear graph passes through $(1, 2)$ and $(3, 4)$.
(o) $(y - 2)^2$ versus x on a log-log graph passes through $(1, 2)$ and $(3, 4)$.
(p) $(y - 2)^2$ versus x on a semilog graph passes through $(1, 2)$ and $(3, 4)$.

8.10 Table 8.9 shows some thermodynamic properties of water. Find the vapor pressure P, liquid and vapor specific volume $\hat{V}$, liquid and vapor specific internal energy $\hat{U}$, and liquid and vapor specific enthalpy $\hat{H}$ at 31.3°C.

8.11 By hand, or using a spreadsheet, plot the power equation $y = 5x^4$ on a rectilinear graph in the range from $x = 0$ to $x = 4.5$.

(a) On the plot with the power equation, show a straight line that passes through $(2, 80)$ and $(4, 1280)$. Through linear interpolation, use this straight line to estimate the value of the power equation at $x = 3$. What is the fractional error and percentage error?

(b) On another graph, plot the power equation in the range from $x = 2.5$ to $x = 3.5$. Show a straight line that passes through $(2.9, 353.64)$ and $(3.1, 461.76)$. Through linear interpolation, use this straight line to estimate the value of the power equation at $x = 3$. What is the fractional error and percentage error?

(c) By comparing the errors in Parts (a) and (b), what do you conclude?

8.12 Write a computer program that allows the user to input an arbitrary number of x and y values that may be correlated by the equations $y = mx + b$ or $y = mx$. The program must ask the user which equation she wants to use to correlate the data. The program

TABLE 8.9
Thermodynamic properties of water

T (°C)	P (bar)	$\hat{V}$ (m³/kg)		$\hat{U}$ (kJ/kg)		$\hat{H}$ (kJ/kg)	
		Liquid	Vapor	Liquid	Vapor	Liquid	Vapor
28	0.0378	0.001004	36.7	117.3	2414.0	117.3	2552.7
30	0.0424	0.001004	32.9	125.7	2416.7	125.7	2556.4
32	0.0475	0.001005	29.6	134.0	2419.4	134.0	2560.0
34	0.0532	0.001006	26.6	142.4	2422.1	142.4	2563.6

which equation she wants to use to correlate the data. The program will calculate the best slope and y-intercept (if there is one) according to least-squares linear regression. The program will also calculate the correlation coefficient. Test the program with the data from Problems 8.1 and 8.2.

8.13 Write a computer program that accomplishes the same task as Problem 8.12, but reads the data from a file rather than having it input directly by the user.

8.14 Write a computer program that linearly interpolates between two data pairs.

Glossary

abscissa   The x-axis.

dependent variable   A variable that cannot be arbitrarily selected and is determined by the independent variable.

exponential equation $y = kB^{mx}$

extrapolation   Extending beyond the data points.

independent variable   A variable that can be arbitrarily selected.

intercept   The value where a line intersects a coordinate axis.

interpolation   Extending between the data points.

least-squares linear regression   A rigorous mathematical procedure that is used to find a line that best fits all the data points.

linear interpolation   The approximation of a curve with a straight line.

linear regression   The process of finding the equation of a straight line that best fits the data.

log-log graph   A graph in which both axes are logarithmic.

ordinate   The y-axis.

power equation $y = kx^m$

rectilinear graph   A graph in which both axes are linear.

regression   Going backward from the data to the equation.

semilog graph   A graph in which one axis is logarithmic and the other is linear.

slope   Rise over run.

vapor pressure   The pressure of the vapor in equilibrium with the liquid or the solid.

CHAPTER 9

SI System of Units

The **SI System of Units** (*Le Système International d'Unités*) is probably familiar to you as the "metric system." This system is used worldwide, although some countries, including the United States, continue to use traditional measurement systems as well. It is the system favored by science, so you undoubtedly have been exposed to it in your science classes.

9.1 HISTORICAL BACKGROUND

The need for units of measure was evident as soon as human commerce began. If two farmers were to trade grain for a goat, they needed to quantify the amount of grain and the weight of the goat. In early commerce, the units of measure were based on commonly available items. For example, the bushel basket used to transport the grain became the unit of measure. (In Britain, the **bushel** was eventually standardized to be eight imperial gallons.) The weight of the goat could be measured by placing the animal on a scale and determining the number of stones required to counterbalance the animal. (In Britain, the **stone** was eventually defined to be 14 pounds.) The unit of length, based on a man's foot, has been used in Britain for over a thousand years. It quickly became evident that units of measure had to be subdivided. Many ancient measuring systems were based upon fractions of the base unit, such as halves, thirds, and quarters. Thus, the unit was subdivided into a number of segments that is easily divided into fractions. For example, the foot is divided into 12 inches, which may be evenly divided by 2, 3, 4, and 6 with no remainder.

For units of measure to be useful, they must be standardized so that business transactions are unambiguous. Thus, it fell upon governments to establish official units of measure. For example, the Egyptian royal **cubit** was equivalent to the length from the Pharaoh's elbow to the farthest fingertip of his extended hand (20.62 inches). A block of granite was fashioned to this length to become a standard. (After all, the Pharaoh was much too busy to help carpenters measure the lengths of boards.) This standard was further divided into finger widths, palms, hands, remens (20 finger widths), and a small cubit (18 inches) equal to six palms (3 inches). The small cubit was used widely in construction and was fashioned into wood or granite copies that were regularly checked against the standard. We continue to use the Pharaoh's system of measure—the height of horses is often measured in **hands,** which are now defined to be exactly four inches.

In the 16th century, decimal systems were conceived in which the units of measure were divided into 10 parts, 100 parts, 1000 parts, and so on, rather than fractional divisions. This allowed for more accurate and convenient subdivisions; however, as there were no standards, confusion abounded. In 1790, the French National Assembly requested that the French Academy of Sciences establish a system of units that could be adopted the world over. It used the meter as the unit of length and the gram as the unit of mass. This system was legalized in the United States in 1866.

In 1870, an international meeting was held in Paris in which 15 nations were represented. This led to the establishment of the International Bureau of Weights and Measures near Paris. They agreed to hold the General Conference on Weights and Measures at least every 6 years to decide upon issues relating to units of measure. The National Institute of Standards and Technology (NIST), formerly the National Bureau of Standards (NBS), represents the United States at these meetings.

Any measuring system must establish **base units** from which all other units are derived. (For example, volume is derived from the base unit of length.) In 1881, time was added as a third base unit to establish the centimeter-gram-second (CGS) system. Outside of the laboratory, this system is inconvenient, so about 1900, the meter-kilogram-second (MKS) was adopted. In 1935, electrical measurements based on the **ampere** were added. Thus there was a fourth base unit in the MKSA system. In 1954, base units for temperature (kelvin) and luminous intensity (**candela**) were adopted, bringing the total base units to six. In 1960, the measurement system was given the formal title *Le Système International d'Unités,* which we abbreviate as SI. In 1971, the amount of substance (mole) was added as a base unit, bringing the total to seven.

9.2 DIMENSIONS AND UNITS

The distinction between a **dimension** and a **unit** is best understood by example. The *dimension* of length may be described by *units* of meters, feet, inches, cubits, and so forth. Thus, *dimension* is an abstract idea whereas *unit* is more specific. Table 9.1 shows common dimensions and the associated **SI base units.**

TABLE 9.1
Dimensions and SI base units

Dimension	Symbol	Unit	Abbreviation
Length	$[L]$	meter	m
Mass	$[M]$	kilogram	kg
Time	$[T]$	second	s
Electric current	$[A]$	ampere	A
Thermodynamic temperature	$[\theta]$	kelvin	K
Luminous intensity	$[I]$	candela	cd
Amount of substance	$[N]$	mole	mol

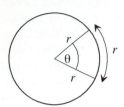

FIGURE 9.1

Plane angle.

9.3   SI UNITS

SI includes three types of units: supplementary, base, and derived.

9.3.1 SI Supplementary Units

The **SI supplementary units** were added in 1960. They are mathematical definitions that are needed to define both base and derived units.

1. Plane Angle (radian).

Figure 9.1 shows a circle in which two radii define a **plane angle** θ. If the length of the swept circumference is equal to the circle radius, then the plane angle θ is equal to one radian (1 **rad**).

Born in Revolution

The erratic behavior and bankruptcy of Louis XVI led to the disintegration of French social order and culminated in the storming of the Bastille on July 14, 1789. This event initiated the destruction of the feudal system as peasants were emboldened to torch chateaux, burn feudal contracts, take land, and drive out the gentry. Political power shifted from the king to the French National Assembly.

In 1790, the French National Assembly requested that the French Academy of Sciences revise the French system of weights and measures, which, under the monarchy, were chaotic, inconsistent, and complex. They appointed a "blue-ribbon" panel headed by Jean Charles Borda (1733–1799). Invitations to sit on the committee were sent to both Britain and the United States, but these were declined.

The committee decided the measuring system should be base 10, although base 11 and 12 were considered. The unit of length, the meter, was to be one 10-millionth of a "quadrant of meridian," that is, the distance from the north pole to the equator measured along a great circle passing through the poles. The unit of mass, the gram, was to be the mass of water, at its maximum density (4°C), occupying a volume of 10^{-6} m^3.

Two French astronomer-geodesists were given the task of surveying the distance from Dunkirk, France, to a site near Barcelona, Spain, from which the quadrant of meridian could be calculated. It took them seven years to complete the measurements; their work was impeded because they were arrested for spying while surveying foreign countries. They were remarkably accurate, with an error of only 2 parts in 10,000.

In 1792, the newly elected National Convention proclaimed France to be a republic. They severed all ties with the traditional Gregorian calendar and established this as Year 1 of the Republic of France. To create a rational calendar, they established a new commission, which devised a 12-month calendar with each month exactly 30 days. To complete the 365-1/4 days in a year, each year had a 5-day festival, except during leap year, which had a 6-day festival. Rather than the traditional 7-day week, the month was divided into three 10-day décades. Rather than naming each day after gods and goddesses, the days were numbered from one to ten. This calendar was employed for over 12 years, until Napoleon abandoned it in 1806.

The calendar commission also proposed a decimal system of time. Each day was divided into ten decidays (2.4 hours); smaller units were the milliday (86.4 seconds) and microday (0.0864 seconds). In 1793, the decimal time system was introduced, but was met with stiff resistance. Unlike the other weights and measures employed by the monarchy, the system for time measurement was rigorous, well structured, and universally followed. Changing all the clocks would be expensive. In addition, time is intimately connected with people's everyday lives, whereas the units for length and mass are less interwoven. There was little incentive for change, so in 1795, the proposed decimal time system was "tabled" and has remained so ever since.

In 1798, European scientists were invited to France to continue improving the new "metric system." Eventually, it became a measuring system that was adopted by nearly the entire world. Interestingly, the United States, which was invited to attend the very first meetings, has had the greatest difficulty adopting this measurement system.

Adapted from: H. A. Klein, *The World of Measurements* (New York: Simon and Schuster, 1974).

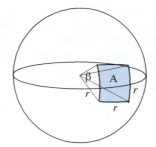

FIGURE 9.2
Solid angle.

Any plane angle θ is defined as the length of the swept circumference divided by the radius:

$$\theta = \frac{\text{swept circumference}}{\text{radius}} = \frac{[\text{L}]}{[\text{L}]} \tag{9-1}$$

Both swept circumference and radius have dimensions of length, so plane angles have dimensions of [L/L]. In SI, the unit for length is the meter (m), so the SI plane angle unit is m/m. These units cancel each other, so plane angles are commonly written without their associated units. Although not formally required, some people prefer to include the abbreviation *rad* after the plane angle.

2. Solid Angle (steradian).

Figure 9.2 shows a sphere in which four radii define a surface with area A. If $A = r^2$, then the solid angle β is equal to one steradian (1 sr).

Any **solid angle** is defined as the swept area divided by the radius squared:

$$\beta = \frac{\text{swept area}}{(\text{radius})^2} = \frac{[\text{L}^2]}{[\text{L}]^2} \tag{9-2}$$

The swept area has dimensions of length squared. Radius has dimensions of length, but because the radius is squared, the denominator also has dimensions of length squared. In SI, the unit for length is the meter (m), so the SI plane angle unit is m^2/m^2. Again, these units cancel each other, so solid angles are commonly written without their associated units. Although not formally required, some people prefer to include the abbreviation *sr* after the solid angle.

9.3.2 SI Base Units

The base units are defined as follows:

1. Unit of Length (meter).

The **meter** was first defined in 1793 by dividing the "quadrant of meridian" (the length from the north pole to the equator measured along a great circle passing through the poles) into 10 million parts. After surveying the earth to determine the length of the quadrant of meridian, the meter was reproduced in three platinum bars and several iron bars. Because of surveying error, it was later found that the bar lengths did not correspond exactly to the original definition. Rather than change the bar lengths, the original definition was abandoned. Because the platinum bars are not easily transported and because they had to be stored at an exact temperature (i.e., the temperature of melting ice) to maintain a given length, they were abandoned as a standard in 1960. Currently, the meter is defined by the distance light traverses in a given length of time.

> *The meter is the length of the path traveled by light in vacuum during a time interval of 1/299792458 of a second.*

2. Unit of Mass (kilogram).

In 1799, the **kilogram** was defined as the mass of pure water at the temperature of its maximum density (4°C) that occupies a cubic decimeter (0.001 m^3). It was later determined that the standard volume used to measure the water was actually 1.000028 cubic decimeters. This definition of the kilogram was abandoned in 1889.

> *The kilogram is defined by a cylindrical prototype composed of an alloy of platinum and 10% iridium maintained under vacuum conditions near Paris.*

The kilogram is the only base unit that is not transportable. Copies are made that match the mass of the original by 1 part in 10^8 or better. Unfortunately, metallurgy in the 19th century was not sophisticated, so impurities in the platinum-iridium cause detectable changes in the prototype mass of about 0.5 part per billion every year. Thus, the definition of the kilogram changes with time.

3. Unit of Time (second).

The unit of time was originally defined as 1/86400 of the mean solar day. Because of irregularities in the earth's rotation, the definition was changed to the "ephemeris **second,**" i.e., 1/31556925.9747 of the tropical year 1900. In 1967, this definition was replaced.

> *The second is the duration of 9,192,631,770 periods of the radiation corresponding to the transition between two hyperfine levels of the ground state of the cesium-133 atom.*

This definition is based on the atomic clock. One of the best atomic clocks (NIST-F1) is precise to within about 1 second in 60 million years, or 5 parts in 10^{16}. Commercially available atomic clocks are precise to within 3 parts in 10^{12}.

4. Unit of Electric Current (ampere).

When electric current flows through a wire, a magnetic field surrounds the wire. The ampere was defined in 1948 on the basis of the magnetic force of attraction between two parallel wires with electric current flowing.

> *The ampere is that constant current which, if maintained in two straight parallel conductors of infinite length, of negligible circular cross section, and placed 1 meter apart in a vacuum, would produce between these conductors a force equal to 2×10^{-7} newton per meter of length.*

This can be better understood by considering Figure 9.3.

5. Unit of Thermodynamic Temperature (kelvin).

Temperature, a measure of random atomic motion, is not to be confused with **heat,** energy flow resulting from a temperature difference. Please review the chapter on thermodynamics if you do not know the difference between the two.

The definition of temperature is based on the **phase diagram** for water (Figure 9.4). The liquid/solid, liquid/vapor, and solid/vapor lines meet at the **triple point,** where all three phases coexist simultaneously. Although it would seem difficult to attain the triple point experimentally because the pressure and temperature combination must be **exact,** it is actually achieved rather easily. A glass vial is evacuated and then partially filled with liquid water, leaving a vapor-space above the liquid. The partially full vial is then frozen. As the ice melts, all three phases will coexist: ice, liquid, and vapor.

Because the triple point of water is rather easily obtained, it is ideal for defining a temperature scale. By definition, the triple point of water is assigned the value 273.16 K and **absolute zero** is assigned the value 0 K. The distance from absolute zero to the triple point of water is divided into 273.16 parts, which define the size of the **kelvin** unit.

The kelvin unit of thermodynamic temperature is the fraction 1/273.16 of the thermodynamic temperature of the triple point of water.

An auxiliary temperature scale is sanctioned in which Celsius temperature t (in °C) is related to Kelvin temperature T (in K) according to the relation

$$t = T - T_o \tag{9-3}$$

where T_o = 273.15 K. In the Celsius temperature scale, water freezes at 0°C and it boils at 100°C provided the pressure is 1 atm. For most engineering work, the Celsius temperature scale is more convenient than the Kelvin temperature scale.

An instrument is needed to divide the interval from absolute zero to the triple point of water, and to extend beyond. In practice, the interval is divided using many different types of instruments (e.g., constant-volume gas thermometers, acoustic gas thermometers, spectral and total radiation thermometers, and electronic noise thermometers). The easiest instrument to understand is the constant-volume gas thermometer. At very low pressures, real gases behave as perfect (ideal) gases. The perfect (ideal) gas equation defines the

FIGURE 9.3
The definition of the ampere.

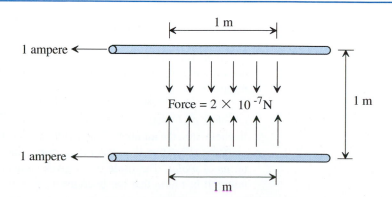

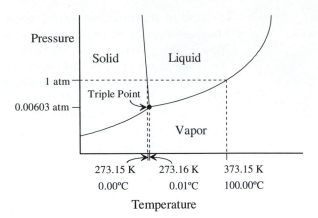

FIGURE 9.4

Phase diagram for water.

relationship between the pressure P, the volume V, the quantity of gas in moles n, and the temperature T,

$$PV = nRT \qquad (9\text{-}4)$$

where R is the universal gas constant. By filling a fixed volume with a given amount of gas, then V and n are constant. The perfect gas equation is simplified to

$$P = \left(\frac{nR}{V}\right)T = kT \qquad (9\text{-}5)$$

where k is the proportionality constant. Thus, pressure is directly proportional to temperature. To illustrate how this relationship could be used, imagine that we perform an experiment in which the pressure in the constant-volume gas thermometer is 0.010000 atm at the triple point of water. If we then reduce the temperature so the pressure in the thermometer becomes 0.0050000 atm, we can calculate the temperature as

$$k = \frac{P_1}{T_1} = \frac{P_2}{T_2} \qquad (9\text{-}6)$$

$$T_2 = \frac{P_2}{P_1}T_1 = \frac{0.0050000 \text{ atm}}{0.010000 \text{ atm}} 273.16 \text{ K} = 136.58 \text{ K} \qquad (9\text{-}7)$$

It is very difficult to make precise and accurate thermometer measurements. The convenient reference points presented in Table 9.2 were determined by very careful thermometry.

6. Unit of Amount of Substance (mole).

In chemistry, the number of molecules is extremely important. For example, the perfect gas equation (Equation 9-4) has the term n, which describes the number of gas molecules in terms of *moles*. The **mole** often gives students difficulty, perhaps because of its unusual name. It is a term that has been used since about 1902 and is short for "gram-**mole**cule."

TABLE 9.2
International Temperature Scale (ITS-90) reference points (Reference)

^{3}He	Boiling point	3.2 K	Ga	Triple point	302.9146 K
^{4}He	Boiling point	4.2 K	In	Freezing point	429.7845 K
H_2	Triple point	13.8033 K	Sn	Freezing point	505.078 K
H_2	Boiling point	20.3 K	Zn	Freezing point	692.677 K
Ne	Triple point	24.5561 K	Al	Freezing point	933.473 K
O_2	Triple point	54.3584 K	Ag	Freezing point	1234.93 K
Ar	Triple point	83.8058 K	Au	Freezing point	1337.33 K
Hg	Triple point	234.3156 K	Cu	Freezing point	1357.77 K
H_2O	Triple point	273.16 K			

Note: Boiling and freezing points measured at $P = 101.325$ kPa

The mole is the amount of substance that contains as many elementary entities as there are atoms in 0.012 kg of carbon-12. When the mole is used, the elementary entities must be specified and may be atoms, molecules, electrons, other particles, or specified groups of such particles.

We can visualize what a mole means by imagining using some submicroscopic tweezers to count the number of atoms in 12 grams (0.012 kg) of carbon-12. The number we obtain is called *Avogadro's number* and is equal to 6.0221367×10^{23}. Just as we use the name *dozen* to describe the number 12, we give a name to this important number.

7. Unit of Luminous Intensity (candela).

A unit for **luminous intensity** is required to describe the brightness of light. Candle flames or incandescent lightbulbs were originally used as standards. The current standard uses a monochromatic (i.e., single-color) light source, typically produced by a laser, and an instrument called a *radiometer* to measure the amount of heat generated when light is absorbed.

The candela is the luminous intensity, in a given direction, of a source that emits monochromatic radiation of a frequency 540×10^{12} cycles per second and that has a radiant intensity in that direction of (1/683) watt per steradian.

Figure 9.5 is a schematic representation of the measuring system for the candela.

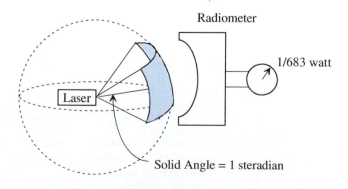

FIGURE 9.5
Schematic representation of the apparatus to measure light intensity.

9.3.3 SI Derived Units

The base units may be combined into the derived units shown in Table 9.3. Some derived units have been assigned special names (Table 9.4). These derived units with special names may even be combined with other units to form new derived units (Table 9.5).

TABLE 9.3
Examples of SI derived units (Reference)

Quantity	SI Unit Name	Symbol
Area	square meter	m^2
Volume	cubic meter	m^3
Speed, velocity	meter per second	m/s
Acceleration	meter per second squared	m/s^2
Wave number	reciprocal meter	m^{-1}
Density, mass density	kilogram per cubic meter	kg/m^3
Specific volume	cubic meter per kilogram	m^3/kg
Current density	ampere per square meter	A/m^2
Magnetic field strength	ampere per meter	A/m
Angular velocity	radian per second	rad/s
Angular acceleration	radian per second squared	rad/s^2
Concentration (amount of substance)	mole per cubic meter	mol/m^3
Luminance	candela per square meter	cd/m^2

TABLE 9.4
SI derived units with special names (Reference)

Quantity	Name	Symbol	Expression in Terms of Other Units	Expression in Terms of SI Base Units
Frequency	hertz	Hz		s^{-1}
Force	newton	N		$m \cdot kg \cdot s^{-2}$
Pressure, stress	pascal	Pa	N/m^2	$m^{-1} \cdot kg \cdot s^{-2}$
Energy, work, heat	joule	J	$N \cdot m$	$m^2 \cdot kg \cdot s^{-2}$
Power, radiant flux	watt	W	J/s	$m^2 \cdot kg \cdot s^{-3}$
Electric charge	coulomb	C		$s \cdot A$
Electric potential	volt	V	W/A	$m^2 \cdot kg \cdot s^{-3} \cdot A^{-1}$
Capacitance	farad	F	C/V	$m^{-2} \cdot kg^{-1} \cdot s^4 \cdot A^2$
Electric resistance	ohm	Ω	V/A	$m^2 \cdot kg \cdot s^{-3} \cdot A^{-2}$
Electric conductance	siemens	S	A/V	$m^{-2} \cdot kg^{-1} \cdot s^3 \cdot A^2$
Magnetic flux	weber	Wb	$V \cdot s$	$m^2 \cdot kg \cdot s^{-2} \cdot A^{-1}$
Magnetic flux density	tesla	T	Wb/m^2	$kg \cdot s^{-2} \cdot A^{-1}$
Inductance	henry	H	Wb/A	$m^2 \cdot kg \cdot s^{-2} \cdot A^{-2}$
Celsius temperature	degree Celsius	°C		K
Luminous flux	lumen	lm		$cd \cdot sr$
Illuminance	lux	lx	lm/m^2	$m^{-2} \cdot cd \cdot sr$
Activity (of a radionuclide)	becquerel	Bq		s^{-1}
Absorbed dose, specific energy imparted	gray	Gy	J/kg	$m^2 \cdot s^{-2}$
Dose equivalent	sievert	Sv	J/kg	$m^2 \cdot s^{-2}$

TABLE 9.5
Examples of SI derived units expressed by means of several names (Reference)

Quantity	SI Unit		Expression in Terms of SI base units
	Name	Symbol	
Dynamic viscosity	pascal second	Pa·s	$m^{-1} \cdot kg \cdot s^{-1}$
Moment of force	newton meter	N·m	$m^2 \cdot kg \cdot s^{-2}$
Surface tension	newton per meter	N/m	$kg \cdot s^{-2}$
Heat flux density, irradiance	watt per square meter	W/m^2	$kg \cdot s^{-3}$
Heat capacity, entropy	joule per kelvin	J/K	$m^2 \cdot kg \cdot s^{-2} \cdot K^{-1}$
Specific heat capacity, specific entropy	joule per kilogram kelvin	J/(kg·K)	$m^2 \cdot s^{-2} \cdot K^{-1}$
Specific energy	joule per kilogram	J/kg	$m^2 \cdot s^{-2}$
Thermal conductivity	watt per meter kelvin	W/(m·K)	$m \cdot kg \cdot s^{-3} \cdot K^{-1}$
Energy density	joule per cubic meter	J/m^3	$m^{-1} \cdot kg \cdot s^{-2}$
Electric field strength	volt per meter	V/m	$m \cdot kg \cdot s^{-3} \cdot A^{-1}$
Electric charge density	coulomb per cubic meter	C/m^3	$m^3 \cdot s \cdot A$
Electric flux density	coulomb per square meter	C/m^2	$m^{-2} \cdot s \cdot A$
Permittivity	farad per meter	F/m	$m^{-3} \cdot kg^{-1} \cdot s^4 \cdot A^2$
Permeability	henry per meter	H/m	$m \cdot kg \cdot s^{-2} \cdot A^{-2}$
Molar energy	joule per mole	J/mol	$m^2 \cdot kg \cdot s^{-2} \cdot mol^{-1}$
Molar entropy, molar heat capacity	joule per mole kelvin	J/(mol·K)	$m^2 \cdot kg \cdot s^{-2} \cdot K^{-1} \cdot mol^{-1}$
Radiant intensity	watt per steradian	W/sr	$m^2 \cdot kg \cdot s^{-3}$
Radiance	watt per square meter steradian	W/(m²·sr)	$kg \cdot s^{-3}$
Exposure (x and γ rays)	coulomb per kilogram	C/kg	$kg^{-1} \cdot s \cdot A$
Absorbed dose rate	gray per second	Gy/s	$m^2 \cdot s^{-3}$

9.4 SI PREFIXES

Because scientists and engineers describe quantities that span many orders of magnitude (e.g., the size of the atomic nucleus to distances between galaxies), SI includes the multipliers listed in Table 9.6. It is generally desirable to use the appropriate multipliers so that the number falls between 0.1 and 1000. (For example, the length 1340 m is best written as 1.34 km.) Exceptions are:

1. In particular applications, a single unit may be customary. For example, engineering mechanical drawings often express all dimensions in millimeters (mm), regardless of how large or small the number is. Clothing dimensions often are expressed in centimeters (cm).

2. When numbers are being compared or listed (as in a table), all numbers should be given with a single prefix.

TABLE 9.6
SI prefixes (Reference)

Multiples			Submultiples		
Factor	Prefix	Symbol	Factor	Prefix	Symbol
10^{24}	yotta	Y	10^{-1}	deci	d[†]
10^{21}	zetta	Z	10^{-2}	centi	c[†]
10^{18}	exa	E	10^{-3}	milli	m
10^{15}	peta	P	10^{-6}	micro	μ
10^{12}	tera	T	10^{-9}	nano	n
10^{9}	giga	G	10^{-12}	pico	p
10^{6}	mega	M	10^{-15}	femto	f
10^{3}	kilo	k	10^{-18}	atto	a
10^{2}	hecto	h[†]	10^{-21}	zepto	z
10^{1}	deka*	da[†]	10^{-24}	yocto	y

* Outside the United States, "deca" is used extensively.
[†] Generally to be avoided.

The use of prefixes eliminates ambiguities associated with significant figures. The number 1340 m could have three or four significant figures, depending on if the last zero is needed merely to place the decimal point. By using the prefix, it is clear that 1.34 km has three significant figures and 1.340 km has four significant figures.

Although prefixes clearly communicate the size of a number, their use in calculations can lead to disasters. It is *strongly* recommended that all numbers in calculations be converted to scientific notation. Thus, if we want to calculate the distance d that light travels in a given time t (say 1 millisecond) given that the speed of light c is 299.8 Mm/s, then we must put these numbers into scientific notation:

$$d = ct \tag{9-8}$$

$$d = \left(299.8 \times 10^6 \frac{m}{s}\right)(1 \times 10^{-3} \, s) = 299.8 \times 10^3 \, m$$

Now that we have the answer, we may wish to communicate it with the appropriate prefix. In this case, the distance could be reported as 299.8 km.

The prefixes that represent 1000 raised to a power are recommended. Thus, the prefixes *hecto*, *deka*, *deci*, and *centi* are generally to be avoided. Some units are so commonly expressed in this way (e.g., centimeter) that their use is accepted.

Use of words such as *billion* and *trillion* are to be avoided in describing multiples of a unit, because the American meaning is different from elsewhere:

Number	United States	Britain, Germany, France
10^{9}	billion	milliard
10^{12}	trillion	billion
10^{15}	quadrillion	—
10^{18}	quintillion	trillion

The Celsius temperature scale was designed to describe temperatures within the range of normal use. Therefore, it is customary not to attach prefixes to the °C symbol. For example, the temperature 5240°C would not be written as 5.24 k°C. For very large (or small) temperatures, it is preferable to use the Kelvin temperature scale.

In the United States, it is customary to express each multiple of 10^3 with the symbol "M." (This notation is derived from the Roman numeral for 1000.) For example, a chemical plant that produces 1,000,000 pounds per year of benzene might be described as a 1 MM lb/year plant. Although this notation is customary, its use should be avoided because of obvious conflicts with SI, in which the prefix "M" means a multiple of 10^6.

9.5 CUSTOMARY UNITS RECOGNIZED BY SI

Table 9.7 shows some customary units that are not formally a part of SI, but are so commonly used that their meaning is regulated by the General Conference on Weights and Measures. Note that the symbol for hour is "h," not "hr" as is commonly used. Also, note that although the symbol for liter is either "l" or "L," the use of the lowercase "l" should be avoided because it is easily confused with the number "1."

Table 9.8 shows some commonly used units that must be experimentally measured. Table 9.9 shows some customary units that are widely used in particular disciplines, but are only temporarily recognized by the General Conference on Weights and Measures. The Conference discourages their introduction into new disciplines.

TABLE 9.7
Customary units recognized by SI (Reference)

Name	Symbol	Value in SI Units
minute of time	min	1 min = 60 s
hour	h	1 h = 60 min = 3600 s
day	d	1 d = 24 h = 86,400 s
degree	°	$1° = (\pi/180)$ rad
minute of arc	′	$1' (1/60°) = (\pi/10,800)$ rad
second of arc	″	$1'' = (1/60') = (\pi/648,000)$ rad
liter	l, L	$1 L = 1 dm^3 = 10^{-3} m^3$
tonne, metric ton	t	$1 t = 10^3$ kg

TABLE 9.8
Experimentally determined units recognized by the SI (Reference)

Name	Symbol	Definition	Measured Value
electron volt	eV	kinetic energy acquired by an electron passing through a potential difference of 1 volt in a vacuum	$1.60217733(49) \times 10^{-19}$ J
unified atomic mass unit	u	(1/12) of the mass of a single atom of ^{12}C	$1.6605402(10) \times 10^{-27}$ kg

Note: The ( ) indicates the uncertainty of the last two significant digits at ± 1 standard deviation.

TABLE 9.9
Units temporarily recognized by SI for use in particular disciplines (Reference)

Name	Discipline or Use	Symbol	Value in SI Units
nautical mile	marine and aerial navigation		1 nautical mile = 1852 m
knot	marine and aerial navigation		1 knot = 1 nautical mile per hour = (1852/3600) m/s
ångström	chemistry, physics	Å	$1\text{ Å} = 0.1\text{ nm} = 10^{-10}\text{ m}$
are	agriculture	a	$1\text{ a} = 1\text{ dam}^2 = 10^2\text{ m}^2$
hectare	agriculture	ha	$1\text{ ha} = 1\text{ hm}^2 = 10^4\text{ m}^2$
barn	nuclear physics	b	$1\text{ b} = 100\text{ fm}^2 = 10^{-28}\text{ m}^2$
bar	meteorology	bar	$1\text{ bar} = 0.1\text{ MPa} = 100\text{ kPa} = 10^5\text{ Pa}$
gal	geodesy and geophysics	Gal	$1\text{ Gal} = 1\text{ cm/s}^2 = 10^{-2}\text{ m/s}^2$
curie	nuclear physics	Ci	$1\text{ Ci} = 3.7 \times 10^{10}\text{ Bq}$
roentgen	exposure to x or γ rays	R	$1\text{ R} = 2.58 \times 10^{-4}\text{ C/kg}$
rad	absorbed dose of ionizing radiation	rad, rd	$1\text{ rad} = 1\text{ cGy} = 10^{-2}\text{ Gy}$
rem	dose equivalent of radiation protection	rem	$1\text{ rem} = 1\text{ cSv} = 10^{-2}\text{ Sv}$

9.6 RULES FOR WRITING SI UNITS (REFERENCE)

1. **Regular upright type (not italics) is used. The *symbol* is written in lowercase except if it was derived from a proper name. The first letter of a symbol derived from a proper name is capitalized.**

    Example: m *is the symbol for* meter *and is written in lowercase letters*
    N *is the symbol for* newton *and is written with an uppercase letter because it originates from the proper name* Newton

    The symbol for liter (L) is an exception, because it was not derived from a proper name. It is capitalized to avoid confusion with the number "1."

2. **The unit *names* are always written in lowercase letters, even if they are derived from a proper name.**

    Example: meter *is the name for the unit of length*
    newton *is the name for the unit, whereas* Newton *is the name of the person*

    An exception is when the unit starts a sentence.

    Example: Newton is the SI unit of force. *Correct*
    newton is the SI unit of force. *Incorrect*

3. **Unit *symbols* are unaltered in the plural (i.e., do not add an "s" to the end of a symbol).**

    Example: The rod length is 3 m. *Correct*
    The rod length is 3 ms. *Incorrect*

    (*Note:* The addition of the "s" to the symbol for meter completely changed the meaning to "millisecond.")

4.   Plurals of the unit *names* are made using the rules of English grammar.

Example: The rod length is 3 meters.   *Correct*
            The rod length is 3 meter.   *Incorrect*

The following units are identical in the singular and plural:

Singular	Plural
lux	lux
hertz	hertz
siemens	siemens

5.   Do not use self-styled abbreviations.

Example: s, A   *Correct*
            sec, amp   *Incorrect*

6.   A space is placed between the symbol and the number.

Example: 5 m   *Correct*
            5m   *Incorrect*

Exceptions are: degrees Celsius (°C), degree (°), minutes ('), and seconds ("), with which there is no space.

Example: 10°C   *Correct*
            10 °C   *Incorrect*

7.   There is no period following the symbol except if it occurs at the end of the sentence.

Example: It took 5 s for the reaction to occur.   *Correct*
            It took 5 s. for the reaction to occur.   *Incorrect*

            The rod is 3 m.   *Correct*
            The rod is 3 m..   *Incorrect*

8.   When a quantity is expressed as a number and unit, and is used as an adjective, then a hyphen separates the number and unit.

Example: The 3-m rod buckled.   *Correct*
            The 3 m rod buckled.   *Incorrect*

            The rod is 3 m.   *Correct*
            The rod is 3-m.   *Incorrect*

9.   The product of two or more unit *symbols* may be indicated with a raised dot or a space.

Example: N·m   *or*   N m

The raised dot is preferred in the United States. Where a raised dot is impossible (e.g., some computer programs), a period may be used instead. An exception is the symbol for watt hour, in which the space or raised dot may be eliminated.

            Wh   *Correct*

10. **The product of two or more unit *names* is indicated by a space (preferred) or a hyphen.**

   Example: newton meter   *or*   newton-meter

   In the case of watt hour, the space may be omitted.

   watthour   *Correct*

11. **A solidus (oblique stroke, /), a horizontal line, or negative exponents may be used to express a derived unit formed from others by division.**

   Example: m/s   *or* $\dfrac{m}{s}$ *or* $m{\cdot}s^{-1}$

12. **The solidus must not be repeated on the same line unless ambiguity can be avoided by parentheses. In complicated cases, negative exponents or parentheses should be used.**

Example: m/s^2 *or* $m{\cdot}s^{-2}$		*Correct*
m/s/s		*Incorrect*
$m{\cdot}kg/(s^3{\cdot}A)$ *or* $m{\cdot}kg{\cdot}s^{-3}{\cdot}A^{-1}$		*Correct*
$m{\cdot}kg/s^3/A$		*Incorrect*

13. **When using the solidus notation, multiple symbols in the denominator must be enclosed in parentheses.**

Example: $m{\cdot}kg/(s^3{\cdot}A)$	*Correct*
$m{\cdot}kg/s^3{\cdot}A$	*Incorrect*

14. **For SI unit names that contain a ratio or quotient, use the word *per* rather than the solidus.**

Example: meters per second	*Correct*
meters/second	*Incorrect*

15. **Powers of units use the modifier *squared* or *cubed* <u>after</u> the unit name.**

Example: meters per second squared	*Correct*
meters per square second	*Incorrect*

   Exceptions are when the unit describes area or volume.

Example: kilograms per cubic meter	*Correct*
kilograms per meter cubed	*Incorrect*

16. **Symbols and unit names should not be mixed in the same expression.**

Example: joules per kilogram *or* J/kg *or* $J{\cdot}kg^{-1}$	*Correct*
joules per kg *or* J/kilogram *or* $J{\cdot}kilogram^{-1}$	*Incorrect*

17. **SI prefix symbols are written in regular upright type (no italics). There is no space or hyphen between the prefix and the unit symbol.**

Example: 5 ms	*Correct*
5 m s	*Incorrect*
5 m-s	*Incorrect*

18. **The entire name of the prefix is attached to the unit name. No space or hyphen separates them.**

Example:	5 milliseconds	*Correct*
	5 milli seconds	*Incorrect*
	5 milli-seconds	*Incorrect*

The final vowel is commonly dropped from the prefix in three cases:

megohm, kilohm, hectare	*Correct*
megaohm, kiloohm, hectoare	*Incorrect*

19. **The grouping formed by the prefix symbol attached to the unit symbol constitutes a new inseparable symbol that can be raised to a positive or negative power and that can be combined with other unit symbols to form compound unit symbols.**

Example:
$$1 \text{ cm}^3 = (10^{-2} \text{ m})^3 = 10^{-6} \text{ m}^3$$
$$1 \text{ cm}^{-1} = (10^{-2} \text{ m})^{-1} = 10^2 \text{ m}^{-1}$$
$$1 \text{ } \mu\text{s}^{-1} = (10^{-6} \text{ s})^{-1} = 10^6 \text{ s}^{-1}$$
$$1 \text{ V/cm} = (1 \text{ V})/(10^{-2} \text{ m}) = 10^2 \text{ V/m}$$

20. **Compound prefixes formed by combining two or more SI prefixes are not permitted.**

Example:	1 mg	*Correct*
	1 μkg	*Incorrect*

Note that even though the kilogram is the base SI unit, multiples are still formed from the gram.

21. **A prefix must have an attached unit and should never be used alone.**

Example:	$10^6/\text{m}^3$	*Correct*
	M/m^3	*Incorrect*

22. **Modifiers are not to be attached to the units.**

Example:	MW of electricity	*Correct*
	MWe	*Incorrect*
	V of alternating current	*Correct*
	Vac	*Incorrect*
	Pa of gage pressure	*Correct*
	Pag	*Incorrect*

If space is limited, the modifier may be placed in parentheses. For example, "Pa (gage)" could be replaced for "Pa of gage pressure."

23. **Use only one prefix in compound units. Normally, the modifier is attached to the numerator.**

Example:	mV/m	*Correct*
	mV/mm	*Incorrect*

An exception is when the kilogram occurs in the denominator.

Example: MJ/kg	*Correct*
kJ/g	*Incorrect*

24. **Dimensionless numbers are not required to have the units reported. For example, the refractive index n is the speed of light in a vacuum c_2 relative to its speed in another medium, c_1.**

$$n = \frac{c_2}{c_1} \tag{9-9}$$

Water has a refractive index of 1.33. It is not necessary to report the units; the same units are used in the numerator (m/s) and denominator (m/s), so they cancel.

In some cases, it is desirable to report the units of dimensionless numbers to avoid confusion. For example, in a mixture containing species A, B, and C, the mass fraction x_A expresses the mass of species A m_A relative to the total mass m_T.

$$x_A = \frac{m_A}{m_T} \tag{9-10}$$

Both the numerator and denominator have units of kilograms, so x_A is a dimensionless number. However, to be absolutely clear, it is best to report the species with the mass.

Example: The mass fraction was 0.1 kg benzene/kg total.	*Preferred*
The mass fraction was 0.1.	*Avoid*

25. **Units such as "parts per thousand" and "parts per million" may be used. However, it is absolutely necessary to explain what the "part" is.**

Example: The mass fraction of CO_2 was 3.1 parts per million.	*Correct*
The mole fraction of CO_2 was 3.1 parts per million.	*Correct*
The fraction of CO_2 was 3.1 parts per million.	*Incorrect*

The adjectives "mass" and "mole" are absolutely essential to clarify the meaning.

26. **Unit symbols are preferred to unit names.**

Example: 15 m	*Preferred*
15 meters	*Avoid*

Many writing conventions require that integers from one to ten be written using words rather than numbers. Therefore, if the number is written in words, then the unit name should be used rather than the symbol.

Example: three meters	*Correct*
three m	*Incorrect*

9.7   SUMMARY

The SI System of Units has three types of units: supplementary, base, and derived. The supplementary units relate to geometry and define the radian and the steradian. There are seven base units: meter, kilogram, second, ampere, kelvin, mole, and candela. Each unit is precisely defined using a transportable standard, except for the kilogram, which still has a prototype standard. The base units may be combined into a variety of derived units, some of which are abbreviated with special names (e.g., Pa for N/m^2).

Because scientists and engineers deal with wide ranges of magnitude, prefixes are employed as multipliers to increase or decrease the size of the units. Great care must be taken when writing units to prevent miscommunication.

Nomenclature

c	speed of light (m/s)
d	distance (m)
k	proportionality constant (atm/K)
m	mass (kg)
n	moles (mol) or refractive index (dimensionless)
P	pressure (atm)
R	universal gas constant (atm·m^3/(mol·K))
T	temperature (K)
t	temperature (°C) or time (s)
V	volume (m^3)
x	mass fraction (dimensionless)
β	solid angle (dimensionless)
θ	plane angle (dimensionless)

Further Readings

ASTM Standard for Metric Practice, E 380-85. Philadelphia: American Society for Testing and Materials, 1989.

The International System of Units (SI), NIST Special Publication 330. United States Department of Commerce, National Institute of Standards and Technology, 1991.

PROBLEMS

9.1 Correct the following units to reflect the proper SI rules:
- **(a)** 18.3 Newton
- **(b)** 45.6 n
- **(c)** 29.0 meter
- **(d)** 56.9 meter/sec
- **(e)** five m
- **(f)** 23 m/second
- **(g)** 493°K
- **(h)** 89.6μ m
- **(i)** 68.5 Kg
- **(j)** 98.4 m/s/s
- **(k)** 10 m's
- **(l)** Mm/ms

9.2 Use an appropriate prefix so the number ranges from 0.1 to 1000:
- **(a)** 9.8×10^5 m
- **(b)** 9.56×10^{10} J
- **(c)** 0.000056 s
- **(d)** 1,984,000 m^3
- **(e)** 35.6×10^{-4} g
- **(f)** 92.4×10^7 N

9.3 A constant-volume gas thermometer is used to measure thermodynamic temperature. At the triple point of water, its pressure is 1.00000×10^2 Pa. What is its pressure at:
- **(a)** triple point of H$_2$
- **(b)** triple point of Ne
- **(c)** triple point of O$_2$
- **(d)** triple point of Ar
- **(e)** triple point of Hg

9.4 Two astronomer-geodesists, J. B. J. Delambre (1749–1822) and P. F. A. Méchain (1744–1804), were appointed to determine the length of a "quadrant of meridian," that is, the length from the north pole to the equator along a great circle passing through the poles. They worked from 1792 to 1799 to complete their task. This length was divided into 10^{-7} parts, the original definition of the meter. In 1799, based upon their measurements, two fine scratches were placed on a platinum bar separated by the distance of one meter. When it was later discovered that there were some small surveying errors, the original definition was abandoned in favor of the platinum bar. For many years, this prototype was the standard that defined the meter.

Later, as measurements improved, the length of the quadrant of meridian was determined to be 10,002,288.3 m. What was the fractional error and percentage error of the measurements made by Delambre and Méchain? If they had access to modern high-precision equipment, would the two fine scratches on the platinum bar have been placed farther apart or closer together? How much longer, or shorter, would the meter be (in millimeters)?

Glossary

absolute zero Temperature is reduced as low as possible and motion, although not at zero, is at its minimum.

ampere The base unit of electric current.

base units The units from which all other units in a measuring system are derived.

bushel A British unit used to measure dry goods; equals 8 imperial gallons.

candela A base unit of luminous intensity.

cubit An ancient form of measure, equivalent to the length from a Pharaoh's elbow to the farthest fingertip of his extended hand.

dimension An abstract idea described by units of measure.

gram-molecule The base unit of the amount of substance (usually referred to as mole); Avogadro's number.

hands A unit of length used to determine the height of a horse; equal to 4 inches.

heat The energy flow resulting from atoms or molecules of different temperatures contacting each other.

kelvin The base unit of thermodynamic temperature.

kilogram The base unit of mass.

luminous intensity The amount of light flux into a specified solid angle.

meter The base unit of length.

mole The base unit of the amount of substance; short for gram-molecule.

phase diagram A diagram that indicates the stable phase or state of a substance at a given temperature and pressure.

plane angle In a circle, the length of the swept circumference divided by the radius.

rad The abbreviation for radian.

SI base units The unit of length (meter), unit of mass (kilogram), unit of time (second), unit of electric current (ampere), unit of thermodynamic temperature (kelvin), unit of amount of substance (mole), and unit of luminous intensity (candela).

SI supplementary units The mathematical definitions needed to define both base and derived units.

SI system of units The metric system.

second The base unit of time.

solid angle In a sphere, the swept area divided by the radius squared.

solidus Oblique stroke: /.

stone A unit of weight in Great Britain; equal to 14 pounds.

temperature A measure of the degree of random atomic motion.

triple point Condition where liquid, solid, and vapor are in equilibrium.

unit A quantity used as a standard of measurement.

CHAPTER 10

Unit Conversions

Mistakes in unit conversions are the most frequent cause of errors in engineering calculations. This is particularly true in the United States, where we employ a mixture of customary and scientific systems. Therefore, the engineer must be well versed in all systems and able to convert among them with ease.

10.1 WHAT DOES IT MEAN TO "MEASURE" SOMETHING?

Whenever we make a measurement, it is always made with respect to a standard. For example, suppose we wished to measure the length of the rod in Figure 10.1. We could approach this problem by obtaining three metersticks and laying them end to end. Because the length of the rod is equivalent to the three metersticks, we report, "the unknown rod is 3 meters in length."

In the United States, we would probably measure the length of the unknown rod using yardsticks. The length of the unknown rod is equivalent to 3 + 0.28 yardsticks, so we report, "the length of the unknown rod is 3.28 yards in length."

FIGURE 10.1

Measuring a rod of unknown length with calibrated sticks.

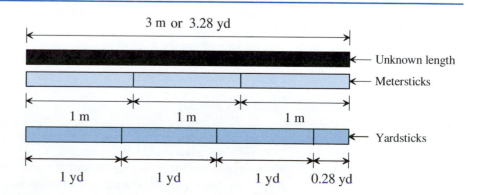

The Right and Lawful Rood

In his 16th-century surveying book, Master Koebel describes how to determine legal lengths. He instructs the surveyor to wait at the church door on Sunday and "bid sixteen men to stop, tall ones, and short ones, as they happen to come out." (Note the use of random selection to prevent bias from having all short, or all tall, men.) These selected men are to stand in line with "their left feet one behind the other"; the accumulated length of their feet defines the "right and lawful rood." The "right and lawful foot" was defined as the sixteenth part of the rood.

Adapted from: H. A. Klein, *The World of Measurements* (New York: Simon and Schuster, 1974), pp. 66–67.

Notice that the reported answer has two parts, the number "3" and the unit "meters" (or "3.28" and the unit "yards"). If we reported "the length of the rod is 3," the reader would not know whether the units were meters, yards, feet, inches, or whatever. *It is absolutely essential to report the unit with the number.* Students often perform elaborate calculations and forget to report the units. Even though the number may be correct, the answer is completely **wrong** without the units.

10.2   CONVERSION FACTORS

When switching between unit systems, it is necessary to use a **conversion factor.** You can find conversion factors in handbook tables and in Appendix A. Conversion factors are developed from identities that are determined experimentally or by definition. For example,

$$1 \text{ ft} \equiv 0.3048 \text{ m}$$

is an exact definition. Because we have this identity where $A = B$, then we know that $A/B = 1$.

$$\frac{1 \text{ ft}}{0.3048 \text{ m}} = 1 = \text{conversion factor} = F$$

$$\frac{0.3048 \text{ m}}{1 \text{ ft}} = 1 = \text{conversion factor} = F$$

Although these conversion factors are not numerically identical to 1, they equal 1 when the units are considered.

We know from the rules of algebra that we are permitted to multiply a quantity by 1 and the quantity remains unchanged. For example, we can multiply the quantity 5 ft by 1 and it remains unchanged.

$$5 \text{ ft} \times 1 = 5 \text{ ft}$$

$$5 \text{ ft} \times F = 5 \text{ ft} \times \frac{0.3048 \text{ m}}{1 \text{ ft}} = 1.524 \text{ m}$$

The units of feet cancel, leaving the unit of meters. Because we have merely multiplied 5 ft by 1, we know that 1.524 m must be identical to 5 ft.

Some students employ an alternative technique:

$$\frac{5 \text{ ft} \quad | \quad 0.3048 \text{ m}}{| \quad 1 \text{ ft}} = 1.524 \text{ m}$$

This method gives the same answer as the other method. It has the advantage of keeping the units neatly organized, but it is less apparent that we are simply multiplying by 1. You may use whichever method you find the most convenient.

A *very* common error occurs when converting a dimension that is raised to a power. Carefully review the following three examples:

$$5 \text{ ft}^2 \times \frac{(0.3048 \text{ m})^2}{(1 \text{ ft})^2} = 0.4676 \text{ m}^2 \qquad\qquad Correct$$

$$5 \text{ ft}^2 \times \left(\frac{0.3048 \text{ m}}{1 \text{ ft}} \right)^2 = 0.4676 \text{ m}^2 \qquad\qquad Correct$$

$$5 \text{ ft}^2 \times \frac{0.3048 \text{ m}^2}{1 \text{ ft}^2} = 1.524 \text{ m}^2 \qquad\qquad Incorrect$$

The first example says that a square that is 1 ft on a side could also be described as a square that is 0.3048 m on a side. The second example says that when the conversion factor F is squared, it is still 1 (i.e., $1^2 = 1$). The third example is completely wrong because the conversion factor for linear dimensions has been confused with the conversion factor for area dimensions.

10.3 MATHEMATICAL RULES GOVERNING DIMENSIONS AND UNITS

In engineering, quantities continually appear in mathematical formulas. The following rules must be obeyed:

Addition/Subtraction: All the terms that are added (or subtracted) must have the same dimensions. For example, if

$$D = A + B - C \tag{10-1}$$

then the dimensions of A, B, C, and D must all be identical.

Multiplication/Division: The dimensions in multiplication/division are treated as though they were variables and cancel accordingly. For example, if A has dimensions of $[M/T^2]$, and B has dimensions of $[T^2/L]$, and C has dimensions of $[M/L^2]$, then D has dimensions of $[L]$ as shown below:

$$D = \frac{AB}{C} = \frac{[M/T^2][T^2/L]}{[M/L^2]} = [L] \tag{10-2}$$

Transcendental Functions: A **transcendental function** cannot be given by algebraic expressions consisting only of the argument and constants. Examples of transcendental functions are

$$A = \sin x \quad B = \ln x \quad C = e^x$$

The argument of a transcendental function (x in these equations) cannot have any dimensions. Similarly, when the transcendental function is evaluated, the result has no dimensions. That is, A, B, and C in these equations have no dimensions.

One can understand the requirement for this rule by realizing that transcendental functions are generally represented by an infinite series. For example, e^x is evaluated with the infinite series:

$$e^x = 1 + x + \frac{x^2}{2!} + \frac{x^3}{3!} + \cdots \tag{10-3}$$

If the argument x had dimensions, then the rule for addition/subtraction would be violated. For example, if x had dimensions [L], then we would be adding [L], [L]^2, [L]^3, and so forth, which we cannot do. Because all the terms on the right must be dimensionless, then clearly e^x must also be dimensionless.

This rule requiring the argument of a transcendental function to be dimensionless applies only to scientific equations with fundamental significance. The rule may be violated in the case of empirical equations. For example, a mechanical engineer interested in the drag force F on an automobile could construct a scale model and test it in a wind tunnel. He would measure the force on the model resulting from different wind velocities v. The data could be correlated with the empirical equation

$$F = av^b \tag{10-4}$$

where a and b are **empirical constants.** This equation can be manipulated by taking the logarithm of both sides:

$$\log F = \log a + \log v^b \tag{10-5}$$

$$\log F = \log a + b \log v \tag{10-6}$$

Plotting $\log F$ versus $\log v$ results in a straight line with a slope b and a y-intercept of $\log a$. Force has dimensions of [ML/T^2] and velocity has dimensions of [L/T], so the argument of the logarithm has dimensions, a violation of the rule. The consequence of this violation is that the constant a will depend on the units chosen to measure force and velocity. For example, the constant a will have one value if force were measured in newtons and velocity in meters per second, and a would have another value if force were measured in poundals (discussed later) and velocity in feet per second.

Dimensional Homogeneity: For an equation to be valid, it must be **dimensionally homogeneous,** that is, the dimensions on the left-hand and right-hand sides of the equation must be the same. For example, Newton's second law

$$F = ma \tag{10-7}$$

$$\left[\frac{\text{ML}}{\text{T}^2} \right] = [\text{M}] \left[\frac{\text{L}}{\text{T}^2} \right]$$

is dimensionally homogeneous because the dimensions on both sides of the equal sign are identical. In contrast, the following equation is not homogeneous,

$$P = \rho T \qquad\qquad \textit{Incorrect} \qquad (10\text{-}8)$$

$$\left[\frac{M}{LT^2}\right] \neq \left[\frac{M}{L^3}\right][\theta]$$

because the dimensions on both sides of the equal sign are not identical.

10.4 SYSTEMS OF UNITS

The SI System of Units, which we discussed in the previous chapter, has evolved in modern times. Many other systems of units have preceded it. Table 10.1 summarizes the most important of these systems. Note that this table divides the unit systems into two major categories: coherent and noncoherent. Within the coherent systems, there are two further subcategories: absolute and gravitational. Coherent/noncoherent and absolute/gravitational systems will be discussed in the next two sections.

Table 10.1 shows some collections of units that are given a name; for example, a $lb_m \cdot ft/s^2$ is called a **poundal**. Be sure to become familiar with the other names given to collections of units.

10.4.1 Absolute and Gravitational Systems of Units

Absolute systems define mass [M], length [L], and time [T]. Force [F] is a derived quantity determined from Newton's second law:

$$\underbrace{[F]}_{\text{Derived}} = F = ma = \underbrace{[M]\frac{[L]}{[T^2]}}_{\text{Defined}} \qquad (10\text{-}9)$$

SI is an absolute system in which mass (kg), length (m), and time (s) are defined, and force (N) is derived.

TABLE 10.1
Systems of units

| Fundamental Dimensions | Coherent | | | | | | Noncoherent |
| | Absolute | | | Gravitational | | | |
	MKS (SI)	CGS	FPS	MKS	CGS	USCS	AES
Length [L]	m	cm	ft	m	cm	ft	ft
Time [T]	s	s	s	s	s	s	s
Mass [M]	kg_m	g_m	lb_m	—	—	—	lb_m
Force [F]	—	—	—	kg_f	g_f	lb_f	lb_f
Derived Dimensions							
Mass [FT²/L]	—	—	—	$kg_f \cdot s^2/m$ (mug)	$g_f \cdot s^2/cm$ (glug)	$lb_f \cdot s^2/ft$ (slug)	—
Force [ML/T²]	$kg_m \cdot m/s^2$ (newton)	$g_m \cdot cm/s^2$ (dyne)	$lb_m \cdot ft/s^2$ (poundal)	—	—	—	—
Energy [LF] or [ML²/T²]	N·m (joule)	dyne·cm (erg)	ft·poundal	$m \cdot kg_f$	$cm \cdot g_f$	$ft \cdot lb_f$	$ft \cdot lb_f$
Power [LF/T] or [ML²/T³]	J/s (watt)	erg/s	ft·poundal/s	$m \cdot kg_f/s$	$cm \cdot g_f/s$	$ft \cdot lb_f/s$	$ft \cdot lb_f/s$

Gravitational systems define force [F], length [L], and time [T]. Mass [M] is a derived quantity, also determined from Newton's second law:

$$[M] = m = \frac{F}{a} = \underbrace{\frac{[F][T^2]}{[L]}}$$

(10-10)

Derived   Defined

Historically, a number of gravitational systems evolved.

Gravitational systems define their units of [F] as pound-force (lb$_f$), kilogram-force (kg$_f$), and gram-force (g$_f$), which are the forces exerted by a pound-mass (lb$_m$), kilogram-mass (kg$_m$), and gram-mass (g$_m$), respectively, under the influence of earth's gravity field (Figure 10.2). Gravity's force can be calculated according to the formula

$$F = mg$$

(10-11)

where g is the acceleration of an object caused by earth's gravity. The value of g is not uniformly the same everywhere on earth. It varies by over 0.5%, depending on the height above sea level and whether extremely dense rock is nearby. Therefore, the **standard** acceleration due to gravity g^o has been defined as

$$g^o = 9.80665 \text{ m/s}^2 = 32.1740 \text{ ft/s}^2$$

(10-12)

(*Note:* The *absolute systems* do not depend on the local strength of the gravity force, hence their name.)

The derived units for mass in the gravitational systems are the **slug, mug,** and **glug** (no joke), derived according to Figure 10.3. The values for the gravitational masses are 1 slug = 32.1740 lb$_m$, 1 mug = 9.80665 kg$_m$, and 1 glug = 980.665 g$_m$. Although the slug is widely used in engineering, the mug and glug are rarely used; they are presented here for the sake of completeness.

Table 10.2 shows some frequently encountered quantities and their associated dimensions in the absolute and gravitational systems. The symbol indicated for each quantity (e.g., "t" for time) is generally encountered in formulas in science and engineering literature, but can change depending on the discretion of the author.

FIGURE 10.2
Gravitational systems. Definitions of pound-force, kilogram-force, and gram-force.

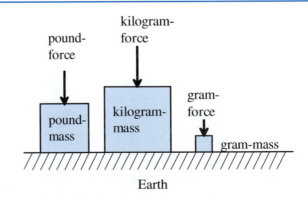

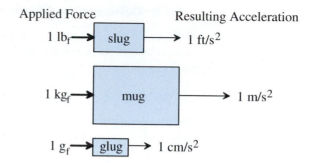

FIGURE 10.3

Gravitational systems. The slug, mug, and glug are derived units.

TABLE 10.2

Frequently encountered engineering quantities (Reference)

Quantity	Symbol	Absolute Dimensions	Gravitational Dimensions
Length	L, l, d	$[L]$	$[L]$
Time	t	$[T]$	$[T]$
Mass	m	$[M]$	$[FT^2/L]$
Force	F	$[ML/T^2]$	$[F]$
Electric current	I, i	$[A]$	$[A]$
Temperature	T	$[\theta]$	$[\theta]$
Amount of substance	n	$[N]$	$[N]$
Luminous intensity	R	$[I]$	$[I]$
Area	A	$[L^2]$	$[L^2]$
Volume	V	$[L^3]$	$[L^3]$
Velocity	v	$[L/T]$	$[L/T]$
Acceleration	a	$[L/T^2]$	$[L/T^2]$
Angular velocity	ω	$[T^{-1}]$	$[T^{-1}]$
Angular acceleration	α	$[T^{-2}]$	$[T^{-2}]$
Frequency	ω	$[T^{-1}]$	$[T^{-1}]$
Electric charge	Q, q	$[AT]$	$[AT]$
Heat capacity	C_p	$[L^2/T^2\theta]$	$[L^2/T^2\theta]$
Kinematic viscosity	ν	$[L^2/T]$	$[L^2/T]$
Momentum	p	$[ML/T]$	$[FT]$
Pressure	P	$[M/LT^2]$	$[F/L^2]$
Stress	σ	$[M/LT^2]$	$[F/L^2]$
Energy (work)	W	$[ML^2/T^2]$	$[FL]$
Energy (heat)	Q, q	$[ML^2/T^2]$	$[FL]$
Torque	τ	$[ML^2/T^2]$	$[FL]$
Power	P	$[ML^2/T^3]$	$[FL/T]$
Density	ρ	$[M/L^3]$	$[FT^2/L^4]$
Dynamic viscosity	μ	$[M/LT]$	$[FT/L^2]$
Thermal conductivity	k	$[ML/T^3\theta]$	$[F/T\theta]$
Voltage	V, E	$[ML^2/AT^3]$	$[FL/AT]$
Electrical resistance	R	$[ML^2/A^2T^3]$	$[FL/A^2T]$

10.4.2 Coherent and Noncoherent Systems of Units

Both the absolute and gravitational systems are **coherent,** meaning that no additional conversion factors are required if the units within that system are used exclusively. For example, the equation

$$F = ma \tag{10-13}$$

may be used in both the FPS (absolute) and USCS (gravitational) systems simply by putting numbers (with their corresponding units) directly into the formula:

$$F = (1 \text{ lb}_m)(10 \text{ ft/s}^2) = 10 \text{ lb}_m{\cdot}\text{ft/s}^2 = 10 \text{ poundal} \qquad \text{FPS}$$

$$F = (1 \text{ slug})(10 \text{ ft/s}^2) = 10 \text{ slug}{\cdot}\text{ft/s}^2 = 10 \text{ lb}_f \qquad \text{USCS}$$

In both cases, the equations are correct; no additional conversion factors were required.

In contrast, the AES (American Engineering System) of units is **noncoherent.** (Some would say it is incoherent, but that's another story.) For example, if we put numbers with their corresponding units into Equation 10-13, *we do not get the correct answer:*

$$F = (1 \text{ lb}_m)(10 \text{ ft/s}^2) \neq 10 \text{ lb}_f \qquad \text{AES}$$

This inconsistency in units results from historical reasons. The pound is a unit that was established in the year 1340. A pound was a pound; there was no distinction between pound-mass and pound-force. It was centuries before Newton taught us the difference between mass, force, and weight.

Mass is the property that causes a body to resist acceleration. A *force* causes a body to accelerate. (The known forces are: gravity, electromagnetism, strong force, and weak force; the latter two are involved with the atomic nucleus.) *Weight* is the force exerted on a body due to local gravity. An object in space can become "weightless" if there is no local gravity. However, it does not lose its mass, because that is a property of the object itself and does not depend on the local environment. In everyday language, the terms *mass* and *weight* are used interchangeably, which adds to the confusion.

Because our forefathers were also confused about the distinction between mass and weight, we have inherited the noncoherent AES system of units. Fortunately, we can overcome the noncoherence by using a conversion factor. This particular conversion factor is given the special name g_c, *but it is still a conversion factor just like the conversion factor F described earlier.* This conversion factor allows us to convert from mass to force or vice versa.

Table 10.3 shows values for g_c in the various systems of units. Notice that the number is unity for all the coherent systems and it differs from unity in the noncoherent system. Because the number is the same as that for the standard acceleration due to gravity, students confuse g^o and g_c. *They are not the same.* The dimensions differ; the dimensions of g^o are [L/T^2] whereas the dimensions of g_c are [ML/T^{2}F]. Students also confuse g_c with g, the local acceleration due to gravity. The conversion factor g_c is a constant that is the same everywhere in the universe (think of it as "g sub constant"), whereas g changes with the local environment. For example, g on the moon is 1/6th that of the earth, but g_c on the moon is unchanged.

TABLE 10.3
The conversion factor g_c

System	Definition		g_c
MKS (absolute)	$1 \text{ N} \equiv 1 \text{ kg}_m \cdot \text{m/s}^2$	$\Rightarrow$	$g_c = 1 \text{ kg}_m \cdot \text{m/(N} \cdot \text{s}^2)$
CGS (absolute)	$1 \text{ dyne} \equiv 1 \text{ g}_m \cdot \text{cm/s}^2$	$\Rightarrow$	$g_c = 1 \text{ g}_m \cdot \text{cm/(dyne} \cdot \text{s}^2)$
FPS (absolute)	$1 \text{ poundal} \equiv 1 \text{ lb}_m \cdot \text{ft/s}^2$	$\Rightarrow$	$g_c = 1 \text{ lb}_m \cdot \text{ft/(poundal} \cdot \text{s}^2)$
MKS (gravitational)	$1 \text{ mug} \equiv 1 \text{ kg}_f \cdot \text{m/s}^2$	$\Rightarrow$	$g_c = 1 \text{ mug} \cdot \text{m/(kg}_f \cdot \text{s}^2)$
CGS (gravitational)	$1 \text{ glug} \equiv 1 \text{ g}_f \cdot \text{cm/s}^2$	$\Rightarrow$	$g_c = 1 \text{ glug} \cdot \text{cm/(g}_f \cdot \text{s}^2)$
USCS (gravitational)	$1 \text{ slug} \equiv 1 \text{ lb}_f \cdot \text{ft/s}^2$	$\Rightarrow$	$g_c = 1 \text{ slug} \cdot \text{ft/(lb}_f \cdot \text{s}^2)$
AES	$1 \text{ lb}_f \equiv 32.174 \text{ lb}_m \cdot \text{ft/s}^2$	$\Rightarrow$	$g_c = 32.174 \text{ lb}_m \cdot \text{ft/(lb}_f \cdot \text{s}^2)$

EXAMPLE 10.1

Problem Statement: What is the weight of 1 lb_m on earth where the local acceleration due to gravity is 32.174 ft/s²? Express the answer in each system of units described in Table 10.3.
Solution: The weight is the force exerted by the object due to gravity. This force can be determined from the equation

$$F = mg \tag{10-14}$$

We place the original units in the equation and then apply appropriate conversion factors from Appendix A to convert the answer to the various systems of units. The factor $1/g_c$ is featured prominently in the conversion.

$$F = (1 \text{ lb}_m)\left(32.174 \frac{\text{ft}}{\text{s}^2}\right)\left(0.4536 \frac{\text{kg}_m}{\text{lb}_m}\right)\left(0.3048 \frac{\text{m}}{\text{ft}}\right)\left(1 \frac{\text{N} \cdot \text{s}^2}{\text{kg}_m \cdot \text{m}}\right) = 4.448 \text{ N} \qquad \text{MKS-abs}$$

$$F = (1 \text{ lb}_m)\left(32.174 \frac{\text{ft}}{\text{s}^2}\right)\left(453.6 \frac{\text{g}_m}{\text{lb}_m}\right)\left(30.48 \frac{\text{cm}}{\text{ft}}\right)\left(1 \frac{\text{dyne} \cdot \text{s}^2}{\text{g}_m \cdot \text{cm}}\right) = 444{,}800 \text{ dyne} \quad \text{CGS-abs}$$

$$F = (1 \text{ lb}_m)\left(32.174 \frac{\text{ft}}{\text{s}^2}\right)\left(1 \frac{\text{poundal} \cdot \text{s}^2}{\text{lb}_m \cdot \text{ft}}\right) = 32.174 \text{ poundal} \qquad \text{FPS-abs}$$

$$F = (1 \text{ lb}_m)\left(32.174 \frac{\text{ft}}{\text{s}^2}\right)\left(0.04625 \frac{\text{mug}}{\text{lb}_m}\right)\left(0.3048 \frac{\text{m}}{\text{ft}}\right)\left(1 \frac{\text{kg}_f \cdot \text{s}^2}{\text{mug} \cdot \text{m}}\right)$$

$$= 0.4536 \text{ kg}_f \qquad \text{MKS-grav}$$

$$F = (1 \text{ lb}_m)\left(32.174 \frac{\text{ft}}{\text{s}^2}\right)\left(0.4625 \frac{\text{glug}}{\text{lb}_m}\right)\left(30.48 \frac{\text{cm}}{\text{ft}}\right)\left(1 \frac{\text{g}_f \cdot \text{s}^2}{\text{glug} \cdot \text{cm}}\right)$$

$$= 453.6 \text{ g}_f \qquad \text{CGS-grav}$$

$$F = (1 \text{ lb}_m)\left(32.174 \frac{\text{ft}}{\text{s}^2}\right)\left(0.03108 \frac{\text{slug}}{\text{lb}_m}\right)\left(1 \frac{\text{lb}_f \cdot \text{s}^2}{\text{slug} \cdot \text{ft}}\right) = 1 \text{ lb}_f \qquad \text{USCS-grav}$$

$$F = (1 \text{ lb}_m)\left(32.174 \frac{\text{ft}}{\text{s}^2}\right)\left(\frac{\text{lb}_f \cdot \text{s}^2}{32.174 \text{ lb}_m \cdot \text{ft}}\right) = 1 \text{ lb}_f \qquad \text{AES}$$

Often in older engineering texts, g_c is incorporated directly into the equation. For example, Equation 10-14 would become

$$F = m \frac{g}{g_c} \tag{10-15}$$

This practice should be avoided because it gives more prominence to g_c than it deserves. After all, none of the other conversion factors were featured in the equation, and they were just as important as g_c. An engineer must always keep track of units; she needs no special reminder about this particular conversion factor.

Another potential source of noncoherence is in the units used to describe energy. For example, the equation for the kinetic energy of an object (i.e., the energy associated with motion) is

$$E = \tfrac{1}{2}mv^2 \tag{10-16}$$

where E is the kinetic energy, m is the mass, and v is the velocity. Energy may be expressed either as "work" (e.g., a force exerted over a distance) or "heat" (energy flow due to a temperature difference). This equation is coherent when energy is expressed as "work" (i.e., joules), but is noncoherent when energy is expressed as heat (e.g., calories).

$$E = \tfrac{1}{2}(1 \text{ kg})\left(2 \ \frac{\text{m}}{\text{s}}\right)^2 = 2 \text{ kg·m}^2/\text{s}^2 = 2 \text{ J} \qquad \textit{Coherent}$$

$$E = \tfrac{1}{2}(1 \text{ kg})\left(2 \ \frac{\text{m}}{\text{s}}\right)^2 = 2 \text{ kg·m}^2/\text{s}^2 \neq 2 \text{ cal} \qquad \textit{Noncoherent}$$

The noncoherent equation, which uses heat to measure energy, can be corrected by using the "mechanical equivalent of heat" (4.1868 J = 1 cal). Thus, by introducing a conversion factor, we can make the noncoherent equation correct:

$$E = \tfrac{1}{2}(1 \text{ kg})\left(2 \ \frac{\text{m}}{\text{s}}\right)^2 = 2 \text{ kg·m}^2/\text{s}^2 = 2 \text{ J} \left(\frac{\text{cal}}{4.1868 \text{ J}}\right) = 0.4777 \text{ cal}$$

It is inconvenient to use this mechanical-equivalent-of-heat conversion factor. Therefore, in SI, we maintain coherence by always expressing heat in joules rather than calories.

10.5   THE DATUM

A **datum** is a reference point used when making a measurement. Suppose we wished to know the difference in height between the third floor and fifth floor of a building (Figure 10.4). We could choose the ground as our *datum*. Because the third floor is 30 ft above the ground and the fifth floor is 50 ft above the ground, we can calculate the difference in height between the third and fifth floors

$$\Delta h = h_5 - h_3 = 50 \text{ ft} - 30 \text{ ft} = 20 \text{ ft}$$

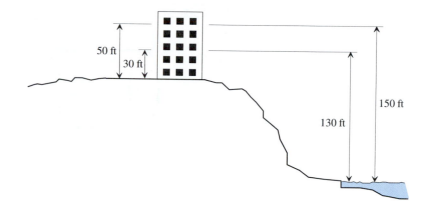

FIGURE 10.4

Illustration of a datum taken as local ground level and another datum taken as sea level.

If we had selected sea level as our datum, we can still calculate the difference in height between the third and fifth floors

$$\Delta H = H_5 - H_3 = 150 \text{ ft} - 130 \text{ ft} = 20 \text{ ft}$$

From this simple example, we see that, provided we are interested in the difference between two numbers, we may select any datum we wish. *The only requirement is that the datum not change in the middle of the calculation.* The use of an arbitrary datum is frequent in science and engineering. For example, the **Celsius temperature scale** uses the freezing point of water as the arbitrary datum.

10.6 PRESSURE

Pressure is a force exerted on an area. **Gas pressure** results from the impact of gas molecules on the container wall, thus exerting a force on a given area of surface (Figure 10.5). If the temperature is increased, the molecules move more vigorously and impact the walls with more force; hence, the pressure increases. For a perfect gas, the perfect (ideal) gas equation describes the relationship between pressure and temperature.

Hydrostatic pressure results from the weight of liquid or gas. If you have ever dived to the bottom of a swimming pool, you have experienced the effects of hydrostatic pressure on your ears. (Each 34 ft of water equals another atmosphere of pressure.) Atmospheric pressure results from the weight of the air above us.

The hydrostatic pressure at the bottom of a container with cross-sectional area A filled with liquid of density ρ can be calculated as follows:

$$P = \frac{F}{A} = \frac{mg}{A} = \frac{\rho V g}{A} = \frac{\rho(Ah)g}{A} = \rho g h \qquad (10\text{-}17)$$

Thus, the hydrostatic pressure depends on the liquid density, the height of the liquid column h, and the local acceleration due to gravity g.

Three types of pressure are generally reported: *absolute, gage,* and *differential.* (A fourth type of pressure, *vacuum pressure,* is discussed later.) We can visualize these by

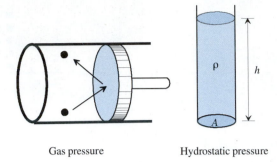

FIGURE 10.5

Illustration of gas pressure and hydrostatic pressure.

Gas pressure　　　　　Hydrostatic pressure

considering Figure 10.6, where the pressure of flowing gas is being measured in a pipe. As the gas flows through the valve, pressure is lost, because some flow energy is converted to heat. The three types of pressures depend on the *datum,* that is, they depend on the reference pressure. In Figure 10.6(a), the reference pressure is a perfect vacuum; the measured pressure is called the **absolute pressure.** In this case, the manometer height h_a indicates an absolute pressure of 25 psia. (*Note:* The "a" refers to absolute.) In Figure 10.6(b), the reference pressure is atmospheric air; the measured pressure is called the **gage pressure.** In this case, the manometer height h_g indicates a gage pressure of 10.3 psig. (*Note:* The "g" refers to gage.) In Figure 10.6(c), an internal reference pressure is used. Because the manometer is measuring the difference between two pressures, it is called **differential pressure.** In this case, the manometer height h_d indicates a differential pressure of 2 psi. (*Note:* There is no need to indicate whether it is absolute or gage, because the difference

FIGURE 10.6

Manometers showing (a) absolute pressure, (b) gage pressure, and (c) differential pressure. (*Note:* Indicated pressures are absolute and in units of psia.)

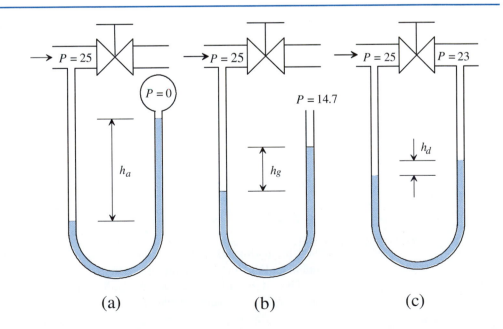

between two readings does not depend on the datum. Some authors would report this as 2 psid, where the "d" refers to "difference.") Although it is common practice to indicate absolute or gage pressures by appending an "a" or "g" to "psi," *this is never done in SI.*

The absolute pressure P_{abs} and gage pressure P_{gage} are related as follows:

$$P_{abs} = P_{gage} + P_{atm} \tag{10-18}$$

where P_{atm} is the atmospheric pressure at the time the gage pressure is measured. The differential pressure, ΔP between two pressures P_1 and P_2 is

$$\Delta P = P_1 - P_2 \tag{10-19}$$

In addition, sometimes pressures are indicated as *vacuum pressure* where larger vacuums are given a larger number. The vacuum pressure P_{vac} is related to absolute pressure as follows:

$$P_{abs} = P_{atm} - P_{vac} \tag{10-20}$$

To ensure that you understand the three common pressure scales (absolute, gauge, and vacuum), closely study Figure 10.7. Notice the datum for each scale, that is, the reference pressure, indicated as zero on each scale. Note also that the pressure difference is the same regardless of the scale.

FIGURE 10.7

Three pressure scales (absolute, gauge, and vacuum) used to indicate the pressure (a) at sea level and (b) on a mountain.

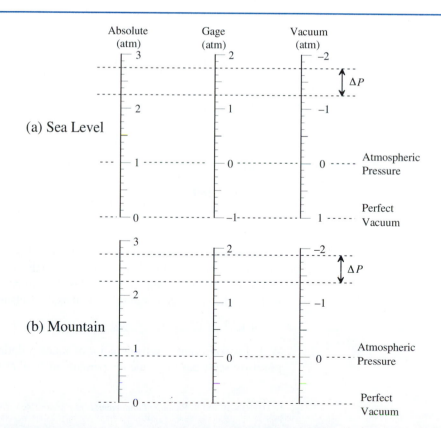

10.7   TEMPERATURE

Temperature is the measure of a body's thermal energy, that is, its molecular (atomic) motion. At high temperatures, molecules (atoms) move vigorously, whereas at low temperatures the motion is less vigorous.

The simplest way to measure temperature is with mercury-in-glass thermometers, which have existed for hundreds of years. At high temperatures, the mercury expands and rises in the thermometer. Once such a thermometer has been constructed, the markings on the glass are arbitrary.

A **temperature scale** is formed by placing two reference points on the mercury-in-glass thermometer and evenly subdividing them into **temperature intervals.** Because students often confuse "temperature scales" and "temperature intervals," here we will establish a nomenclature so the two are easily distinguished. A "temperature scale" will be indicated by placing the degree sign first (e.g., °C) whereas a "temperature interval" will be indicated by placing the degree sign last (e.g., C°). (*Note:* This convention is used in Halliday and Resnick, *Fundamentals of Physics,*[1] but it is rarely used in the general literature. In fact, SI removed the degree symbol from the Kelvin temperature scale in 1967, so it is simply "K," not "°K." However, here we will leave the degree symbol and manipulate it to suit our instructional purposes.)

Around 1714, G. D. Fahrenheit (1686–1736) developed the **Fahrenheit temperature scale.** The two reference points were the freezing point of water (32°F) and the boiling point of water (212°F). (The temperature 0°F corresponds to the coldest temperature achievable with water, salt, and ice.)

In 1743, Anders Celsius (1701–1744) devised the **Celsius temperature scale.** The two reference points were the melting point of ice (0°C) and the boiling point of water (100°C). It was originally known as the "centigrade scale" ("one hundred steps"), but this name has been abandoned in favor of the official SI name, Celsius.

The Fahrenheit scale is almost twice as sensitive as the Celsius scale; 180 F° span the melting point to the boiling point, whereas 100 C° span this same interval. Thus, we can find the relationship between the Fahrenheit interval (F°) and the Celsius interval (C°) as

$$100 \text{ C}° = 180 \text{ F}° \tag{10-21}$$

$$1 \text{ C}° = \frac{180}{100} \text{ F}° = \frac{9}{5} \text{ F}° = 1.8 \text{ F}° \tag{10-22}$$

Rather than using the arbitrary reference points of the Celsius and Fahrenheit scales, **thermodynamic temperature scales** (also called *absolute temperature scales*) specify that the temperature should be proportional to the thermal energy of an ideal gas. There are two thermodynamic temperature scales: Kelvin and Rankine. The **Kelvin temperature scale** has the same temperature interval as the Celsius scale:

$$1 \text{ K}° = 1 \text{ C}° \tag{10-23}$$

With this constraint, the triple point of water is defined as 273.16°K. The **Rankine temperature scale** has the same temperature interval as the Fahrenheit temperature scale

---

1. D. Halliday and R. Resnick, *Fundamentals of Physics* (New York: John Wiley & Sons, 1974) p. 351.

$$1 \, R^\circ = 1 \, F^\circ \tag{10-24}$$

With this constraint, the triple point of water is defined as 491.69°R. The thermodynamic temperature scales are named after J. K. Rankine (1820–1872) and Lord Kelvin (1824–1907), who helped establish them.

Figure 10.8 shows the four temperature scales: Celsius, Kelvin, Fahrenheit, and Rankine. The relationships between these various scales follow:

$$[^\circ R] = [^\circ F] + 459.67 \tag{10-25}$$

$$[^\circ K] = [^\circ C] + 273.15 \tag{10-26}$$

$$[^\circ C] = \frac{[^\circ F] - 32}{1.8} \tag{10-27}$$

$$[^\circ F] = 1.8 \, [^\circ C] + 32 \tag{10-28}$$

$$[^\circ R] = 1.8 \, [^\circ K] \tag{10-29}$$

$$[^\circ K] = \frac{[^\circ R]}{1.8} \tag{10-30}$$

where [°R] is a Rankine temperature, [°F] is a Fahrenheit temperature, [°K] is a Kelvin temperature, and [°C] is a Celsius temperature.

The conversion factors for the temperature intervals are

$$\text{Conversion factor} = \frac{1.8 \, F^\circ}{C^\circ} = \frac{1.8 \, R^\circ}{K^\circ} = \frac{1 \, F^\circ}{R^\circ} = \frac{1 \, C^\circ}{K^\circ} = \frac{1.8 \, F^\circ}{K^\circ} = \frac{1.8 \, R^\circ}{C^\circ} \tag{10-31}$$

FIGURE 10.8

The four temperature scales. (*Note:* Temperature scales are indicated by placing the degree symbol first, whereas temperature intervals are indicated by placing the degree symbol last. According to SI conventions, the degree symbol has been eliminated from the Kelvin scale, but here it is included for instructional purposes.)

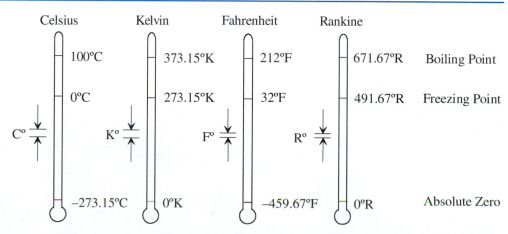

EXAMPLE 10.2

Problem Statement: Calculate the heat transfer Q/t through a block that has a cross-sectional area A of 1 ft^2, a length x of 2 ft, and a thermal conductivity k of 5 Btu/(h·ft·F°). One surface is maintained at 200°F and the other is maintained at 150°F (Figure 10.9). Do the calculation in both U.S. engineering units and SI units.

Solution: The equation for conductive heat transfer follows:

$$\frac{Q}{t} = k\frac{A\Delta T}{x} \tag{10-32}$$

The temperature difference ΔT is a temperature interval calculated as follows:

$$\Delta T = 200°F - 150°F = 50\ F°$$

We may put this into Equation 10-32 to calculate the heat transfer Q/t,

$$\frac{Q}{t} = \left(5\ \frac{\text{Btu}}{\text{h·ft·F°}}\right)\frac{(1\ \text{ft}^2)(50\ F°)}{2\ \text{ft}} = 125\ \text{Btu/h}$$

We can convert all the given information to SI:

$$\Delta T = 50\ F°\left(\frac{K°}{1.8\ F°}\right) = 27.8\ K°$$

$$k = 5\ \frac{\text{Btu}}{\text{h·ft·F°}}\left(\frac{\text{h}}{3600\ \text{s}}\right)\left(\frac{\text{ft}}{0.3048\ \text{m}}\right)\left(\frac{1055\ \text{J}}{\text{Btu}}\right)\left(\frac{1.8\ F°}{K°}\right) = 8.65\ \frac{\text{J}}{\text{s·m·K°}}$$

$$A = 1\ \text{ft}^2\left(\frac{0.3048\ \text{m}}{\text{ft}}\right)^2 = 0.0929\ \text{m}^2$$

$$x = 2\ \text{ft}\left(\frac{0.3048\ \text{m}}{\text{ft}}\right) = 0.610\ \text{m}$$

FIGURE 10.9

Heat transfer through a block.

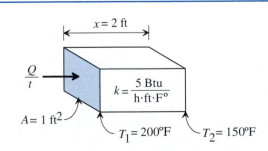

These SI quantities may now be placed in Equation 10-32.

$$\frac{Q}{t} = \left(8.65 \frac{\text{J}}{\text{s·m·K°}} \right) \frac{(0.0929 \text{ m}^2)(27.8 \text{ K°})}{(0.610 \text{ m})} = 36.6 \text{ J/s} = 36.6 \text{ W}$$

The conversion factors in Appendix A show that these two answers (125 Btu/h and 36.6 W) are identical, as they must be.

10.8   CHANGING THE SYSTEM OF UNITS IN AN EQUATION

Engineers must often convert equations from one system of units to another. For example, the perfect gas equation

$$PV = nRT \qquad\qquad (10\text{-}33)$$

is learned in chemistry classes but is widely used in engineering as well. The probable units used in each discipline are shown here:

Quantity	Chemistry (SI)	Engineering
P	Pa	psia
V	m^3	ft^3
n	mol	lbmol
T	K	°R
R	Pa·m^3/(mol·K)	psia·ft^3/(lbmol·°R)

The perfect gas constant R is empirically measured and contains the necessary units to make the equation dimensionally homogeneous (i.e., it makes the units to the left and right of the equal sign the same).

EXAMPLE 10.3

Problem Statement: Convert the SI perfect gas equation

$$PV = n \left(8.314 \frac{\text{Pa·m}^3}{\text{mol·K}} \right) T$$

into an equation in which engineering units may be used for the variables.

Solution: Because the perfect gas constant R contains the necessary units to make the equation dimensionally homogeneous, we merely have to change the units of R to engineering units:

$$R = \left(8.314 \frac{\text{Pa·m}^3}{\text{mol·K}} \right) \left(\frac{\text{psia}}{6895 \text{ Pa}} \right) \left(\frac{35.31 \text{ ft}^3}{\text{m}^3} \right) \left(\frac{453.6 \text{ mol}}{\text{lbmol}} \right) \left(\frac{\text{K}}{1.8°\text{R}} \right)$$

$$= 10.73 \frac{\text{psia·ft}^3}{\text{lbmol·°R}}$$

The perfect gas equation in engineering units now becomes

$$PV = n\left(10.73 \; \frac{\text{psia·ft}^3}{\text{lbmol·°R}}\right)T$$

10.9 DIMENSIONAL ANALYSIS (ADVANCED TOPIC)

Dimensional analysis is widely used in engineering to solve problems about which there is little fundamental information. Simply by looking at the dimensions of the quantities involved, we can tell much about how the quantities are related.

As an example, consider the pendulum shown in Figure 10.10. In the design of a grandfather clock, it would be very useful to know the *period,* that is, the length of time it takes for the pendulum to swing back and forth. Before we perform a sophisticated analysis of the problem, we can use our intuition and experience to simply write down the relevant quantities and their dimensions:

p	period [T]
m	mass [M]
g	acceleration due to gravity [L/T^2]
L	length [L]

We propose that the period is proportional to these quantities according to the following equation:

$$p = km^a g^b L^c \qquad\qquad (10\text{-}34)$$

Using dimensional analysis, we can determine the exponents required to make the equation dimensionally homogeneous. We consider each dimension one at a time.

	[T]	=	[M]^a		[L/T^2]^b		[L]^c		
[M]	0	=	a	+	0	+	0	$\Rightarrow$	$a = 0$
[T]	1	=	0	−	$2b$	+	0	$\Rightarrow$	$b = -1/2$
[L]	0	=	0	+	b	+	c	$\Rightarrow$	$c = +1/2$

FIGURE 10.10
Pendulum.

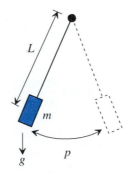

In the first row, [M] appears only on the right side. Therefore, the exponent a must be zero. In the second row, [T] appears to the first power on the left and to the negative two power on the right. Therefore, the exponent b must be $-1/2$. In the third row, [L] does not appear on the left, so the lengths on the right must cancel. Because we have already determined that b is $-1/2$, then c must be $+1/2$.

Using dimensional analysis, we know that p must be proportional to $g^{-1/2}$ and $L^{+1/2}$:

$$p \propto g^{-1/2} L^{+1/2} = kg^{-1/2} L^{+1/2} = k\sqrt{\frac{L}{g}} \qquad (10\text{-}35)$$

Although this simple analysis does not give the value of k, we have learned much about the system. Surprisingly, the period does not depend on the mass.

Dimensional analysis may be applied to fluid flow in a pipe (Figure 10.11). The pressure drop ΔP (i.e., $P_1 - P_2$) is needed to specify the pump size. The relevant quantities for this problem are

ΔP	pressure [M/LT^2]
D	pipe diameter [L]
L	pipe length [L]
v_{ave}	average fluid velocity in pipe [L/T]
ρ	fluid density [M/L^3]
μ	fluid dynamic viscosity [M/LT]

Dimensional analysis has shown that these quantities can be grouped into the following equation:

$$\left(\frac{\Delta P D}{2L\rho v_{ave}^2}\right) = k\left(\frac{D v_{ave} \rho}{\mu}\right)^a \qquad (10\text{-}36)$$

Both terms in brackets are dimensionless. The left one is the *Fanning friction factor f* and the right one is the *Reynolds number* Re. Therefore, Equation 10-36 may be written as

$$f = k(\text{Re})^a \qquad (10\text{-}37)$$

Experimentation with many fluids (e.g., air, water, oil) has shown that $k = 16$ and $a = -1$ provided Re < 2300.

We often encounter dimensionless groups in engineering because many systems are too complicated to be analyzed using mechanistic models. Instead, we write down the relevant quantities, make dimensionless groups using dimensional analysis, and then perform

FIGURE 10.11

Pressure drop of fluid flowing through a pipe.

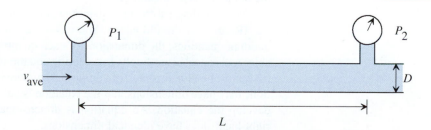

TABLE 10.4

Some common dimensionless groups (Reference).

Name	Symbol	Definition
Fanning friction factor	f	$\dfrac{\Delta P D}{2 L \rho (v_{ave})^2}$
Reynolds number	Re	$\dfrac{D v_{ave} \rho}{\mu}$
Drag coefficient	C_d	$\dfrac{2F}{\rho A v^2}$
Prandtl number	Pr	$\dfrac{C_p \mu}{k}$
Nusselt number	Nu	$\dfrac{hD}{k}$
Mach number	Ma	$\dfrac{v}{a}$

ΔP = pressure drop D = diameter L = length
ρ = density v_{ave} = average velocity μ = viscosity
C_p = heat capacity k = thermal conductivity h = heat transfer coeff.
v = speed a = speed of sound F = force
A = projected area

experiments to determine the value of the proportionality constant and the exponent(s). Table 10.4 shows some dimensionless groups that frequently occur in engineering. Many of these dimensionless groups have fundamental significance. For example, the Reynolds number gives the inertia force/viscous force.

10.10   SUMMARY

Because there are so many unit systems in use, engineers must be familiar with a wide variety of conversions. Mistakes with unit conversions are among the most common errors made by engineers.

When reporting a measurement or computed answer, it is absolutely essential to report the units along with the number. The units indicate the reference; without the reference, the number has no meaning.

When adding/subtracting quantities, the dimensions of each quantity must be identical. (Remember the old adage: You cannot add apples and oranges.) When multiplying/dividing quantities, the dimensions of each quantity are treated as though they were variables and cancel accordingly. In scientific equations, the arguments of transcendental functions must be dimensionless. In empirical equations, the arguments of transcendental functions may have dimensions, but the equation is valid only if you use the units employed to develop the equation. An equation is dimensionally homogeneous if the left-hand and right-hand sides have identical dimensions.

Why Do Golf Balls Have Dimples?

Experiments with smooth golf balls and dimpled golf balls reveal that those with dimples travel much faster and farther. As a smooth golf ball travels through the air, there is a vacuum created in its wake because the air is not able to quickly fill in the void behind

lessening the vacuum and allowing the golf ball to travel faster and farther.

An engineer designing a golf ball is interested in the drag force as a function of velocity. Using dimensionless groups, this

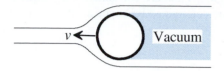 

the golf ball. The vacuum behind the golf ball slows it down and prevents it from traveling great distances. A dimpled golf ball also has a vacuum in its wake, but it is much less. The turbulent vortices caused by the dimples force some air behind the golf ball, thus

would be quantified by plotting the drag coefficient versus the Reynolds number. In the region of Reynolds numbers (i.e., velocities) achievable with a golf club, the drag coefficient is much less with a dimpled ball.

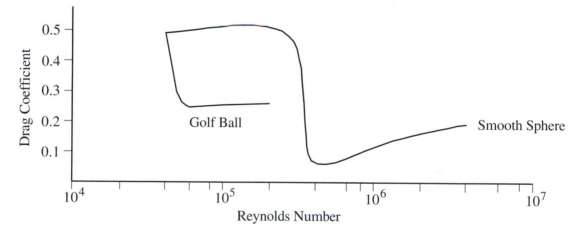

Adapted from: P. Moin and J. Kim, "Tackling Turbulence with Supercomputers," *Scientific American* 276, no. 1 (1997), pp. 62–68.

In coherent unit systems, the numerical part of any conversion factor is 1; however, in noncoherent unit systems, the numerical part of conversion factors differs from 1. For example, in coherent systems, the numerical part of g_c is 1 [e.g., 1 kg·m/(N·s^2)], but in noncoherent unit systems, the numerical part of g_c is not 1 [e.g., 32.174 lb$_m$·ft/(lb$_f$·s^2)]. There are two types of coherent unit systems: gravitational and absolute. Gravitational systems define force, length, and time; mass is derived. Absolute unit systems define mass, length, and time; force is derived. SI is an example of an absolute system.

A datum is a reference point used in measurement. When measuring pressure, the datum can be a perfect vacuum (absolute pressure), atmospheric pressure (gauge pressure), or an internal reference (pressure difference). When measuring temperature, the datum can be absolute zero (Rankine and Kelvin temperature scales), the freezing point of water (Celsius temperature scale), or the coldest temperature achievable using water, ice, and salt (Fahrenheit temperature scale).

Dimensional analysis is a process of determining the functional relationship between quantities by using the dimensions of each quantity. It is commonly used in complex phenomena that defy mechanistic modeling, but are amenable to experiment. Typically, the quantities are arranged into dimensionless groups that are correlated using a power equation. The constants in the power equation are determined by experiment.

Nomenclature

A area (m^2)
a acceleration (m/s^2) or proportionality constant ($N \cdot s^c/m^c$) or speed of sound (m/s)
b exponent (dimensionless)
C_p constant-pressure heat capacity [$J/(kg \cdot K)$]
D diameter (m)
E energy (J)
F force (N)
g acceleration due to gravity (m/s^2)
k heat transfer coefficient [$J/(s \cdot m \cdot K)$] or proportionality constant
H height (m)
h height (m) or heat transfer coefficient [$J/(s \cdot m^2 \cdot K)$]
L length (m)
m mass (kg)
n mole (mol)
P pressure (Pa)
p period (s)
Q heat (J)
R universal gas constant [$Pa \cdot m^3/(mol \cdot K)$]
T temperature (K)
t time (s)
V volume (m^3)
v velocity (m/s)
x spacing (m)
μ viscosity [$kg/(m \cdot s)$]
ρ density (kg/m^3)

Further Readings

Eide, A. R., R. D. Jenison, L. L. Northup, and S. K. Mikelson *Engineering Fundamentals and Problem Solving,* 5th ed. New York: McGraw-Hill, 2008.

Felder, R. M., and R. W. Rousseau. *Elementary Principles of Chemical Processes,* 3rd ed. New York: Wiley, 1999.

Jerrard, H. G., and D. B. McNeill. *A Dictionary of Scientific Units,* 6th ed. rev. Netherlands: Kluwer Academic Publishers, 1992.

PROBLEMS

Note: These problems are solved using the unit conversions in Appendix A.

10.1 Convert the following numbers to SI:
(a)   56.8 in
(b)   1 year
(c)   34.8 atm
(d)   8.3 Btu/min
(e)   38.96 ft/min
(f)   98.6 furlongs per fortnight
(g)   34.5 in·pdl
(h)   15.8 gal/min

10.2 Perform the following conversions:
(a)   34,589 Pa to psia
(b)   34.6 m/s to mph
(c)   89.6 L to m³
(d)   68.4 in³ to ft³
(e)   964 slug to lb_m
(f)   569 pdl to lb_f
(g)   56.8 glug to g
(h)   78.5 mug to kg
(i)   358 atm·L to Btu
(j)   1.2 steradians to square degrees

10.3 Convert the following pressures to absolute pressure. Report the answer in the given units. You may assume that the atmospheric pressure at the time of the measurement was a standard atmosphere.
(a)   8.56 psig
(b)   18 inches of mercury vacuum
(c)   10 atm gage pressure
(d)   15.6 inches of water gage pressure

10.4 Convert the following temperatures to the Kelvin temperature scale:
(a)   259.6°R
(b)   145°C
(c)   98.6°F

10.5 Convert the following temperature intervals to Kelvin.
(a)   58.3 C°
(b)   89.7 F°
(c)   4.5 R°

10.6 A number of objects are taken to the moon, where the acceleration due to gravity is about 1/6 of earth's. Complete the following table:

Item	Mass on Earth	Mass on Moon	Weight on Moon	Weight on Moon
Paper clip	1.2 g	g	g_f	dyne
Can of cola	0.56 lb_m	lb_m	lb_f	pdl
Hammer	1.3 kg	kg	kg_f	N

10.7 Are the following equations dimensionally homogeneous? (Show how you made your conclusion.)
(a) $F = PT$ F = force, P = pressure, T = temperature
(b) $E = 0.5mv^2$ E = energy, m = mass, v = velocity
(c) $p = mv$ p = momentum, m = mass, v = velocity
(d) $E = C_p/P$ E = energy, C_p = heat capacity, P = pressure
(e) $V = IR$ V = voltage, I = current, R = resistance
(f) $P = I^2R$ P = power, I = current, R = resistance

10.8 Compute the ideal gas constant R using the following units of measure:

P = inches of mercury
V = gal
n = short-ton moles
T = Rankine

10.9 A weir is a slot in an open channel with water flowing over it (Figure 10.12). For a weir with a rectangular shape, the following formula can be derived:

$$Q = 5.35Lh^{3/2}$$

where

Q = water flow rate, ft³/s
L = length of weir, ft
h = height of liquid above weir gate, ft

Do the following:
(a)   Determine the units associated with the constant that are needed to make the equation dimensionally homogeneous.
(b)   Determine a new constant that allows the following units to be used:

Q = water flow rate, gal/min
L = length of weir, ft
h = height of liquid above weir gate, in

10.10 Compute the Reynolds number defined in Table 10.4.
(a)   Use the following information:
v_{ave} = 0.3 ft/min

FIGURE 10.12

Water flowing over a weir.

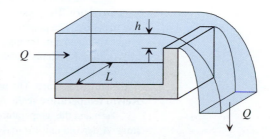

$D = 2$ in
$\rho = 62.3$ lb$_m$/ft^3
$\mu = 2.42$ lb$_m$/(ft·h)

- **(b)** Use the equivalent SI units to the data given in Part (a).
- **(c)** Use the following information:

$v_{ave} = 0.3$ ft/min
$D = 2$ in
$\rho = 62.3$ lb$_m$/ft^3
$\mu = 2.09 \times 10^{-5}$ lb$_f$·s/ft^2

10.11 Determine the pressure difference between the following two pressures (use the units given). You may assume that the atmospheric pressure was one standard atmosphere when the pressure measurements were made.

- **(a)** $P_1 = 1.5$ atm (absolute), $P_2 = 1.3$ atm (absolute)
- **(b)** $P_1 = 1.5$ atm (absolute), $P_2 = 0.3$ atm (gage)
- **(c)** $P_1 = 1.24$ psig, $P_2 = 1.00$ psig
- **(d)** $P_1 = 10.3$ psig, $P_2 = 16.3$ psia

10.12 Decide whether you would use a mercury manometer (at 0°C) or a water manometer (at 4°C) to measure the following pressures.

(*Note:* Each manometer will use a perfect vacuum as a reference. At 0°C, mercury has a negligible vapor pressure. At 4°C, water has a vapor pressure of 813 Pa.)

- **(a)** 34 Pa (absolute)
- **(b)** 1.45 psig
- **(c)** 0.75 atm (absolute)
- **(d)** 1.5 millibar (absolute)

10.13 Write a spreadsheet that converts lb$_m$ to kg, ft to m, in to m, ft^2 to m^2, gal to m^3, and ft^3 to m^3.

10.14 Using Equation 10-37, write a computer program that calculates the pressure drop (in psi) of water flowing through a 1-mile pipe. Use the water properties and pipe diameter given in Problem 10.10. Use v_{ave} increments of 0.1 ft/min until the Reynolds number exceeds 2300, the upper limit where the equation is valid.

10.15 Perform the task described in Problem 10.14 using a spreadsheet. In addition, plot the pressure drop (in psi) versus the average velocity (in ft/min).

Glossary

absolute pressure The pressure relative to a perfect vacuum.
absolute systems A coherent system that defines mass, length, and time; force is a derived quantity.
Celsius temperature scale A temperature scale in which the two reference points are the melting point of ice (0°C) and the boiling point of water (100°C).
coherent unit system A unit system in which the numerical part of any conversion factor is 1.
conversion factor A numerical factor used to multiply or divide a quantity when converting one unit system to another.
datum A reference point used when making a measurement.
differential pressure The difference between two pressures.
dimensional analysis A process of determining the functional relationship between quantities by using the dimensions of each quantity.
dimensional homogeneity The dimensions on the left-hand and right-hand sides of the equation must be the same.
empirical constant The value of the constant is determined by experiment.
Fahrenheit temperature scale A temperature scale in which the two reference points are the freezing point of water (32°F) and the boiling point of water (212°F).
gage pressure The pressure relative to atmospheric air.
gas pressure The pressure resulting from the impact of gas molecules on the container wall.
glug A derived unit for mass in the CGS gravitational system; rarely used.
gravitational systems A coherent system that defines force, length, and time. Mass is a derived quantity.
hydrostatic pressure The pressure resulting from the weight of liquid or gas.
Kelvin temperature scale A thermodynamic temperature scale in which 0 K is defined as absolute zero and the temperature interval is the same as the Celsius scale.
mug A derived unit for mass in the MKS gravitational system; rarely used.

noncoherent unit system A unit system in which the numerical part of conversion factors differs from 1.

poundal A unit of force in the FPS absolute system.

pressure A force exerted on an area.

Rankine temperature scale A thermodynamic temperature scale in which 0°R is defined as absolute zero and the temperature interval is the same as the Fahrenheit scale.

slug A derived unit for mass in the USCS gravitational systems; widely used in engineering.

temperature interval The difference between two temperatures.

temperature scale Formed by placing two reference points on the mercury-in-glass thermometer and evenly subdividing the points into temperature intervals.

thermodynamic temperature scales Temperature scale that is linearly proportional to the thermal energy of an ideal gas; also called absolute temperature scales.

transcendental functions Mathematical functions represented by an infinite series.

APPENDIXES

APPENDIXES

APPENDIX A

UNIT CONVERSION FACTORS

A.1 PLANE ANGLE

$$\theta = \frac{x}{r} = \frac{[L]}{[L]}$$

TABLE A.1
Plane angle conversion factors (Reference)

	°	′	″	rad	rev
1 degree =	1	60	3600	$\pi/180$	1/360
1 minute =	1/60	1	60	$\pi/10,800$	1/21,600
1 second =	1/3600	1/60	1	$\pi/648,000$	1/1,296,000
1 radian =	$180/\pi$	$10,800/\pi$	$648,000/\pi$	1	$1/(2\pi)$
1 revolution =	360	21,600	1,296,000	2π	1

90° = 100 grade [a] = 100^g = 100 gon   90° = 1000 angular mil [b]

[a] All grade subdivisions are indicated with decimals, so there are no equivalent units of minutes or seconds. This system is not widely used except in France.

[b] During World War II, the U.S. artillery divided a right angle into 1000 parts called *angular mil*.

An angle θ is defined by

$$\theta \equiv \frac{x}{r} \tag{A-1}$$

where the angle is measured in radians. Because the perimeter around a circle is $2\pi r$, one complete revolution is

$$\theta = \frac{2\pi r}{r} = 2\pi \left(\frac{r}{r}\right) = 2\pi \text{ rad} \tag{A-2}$$

The perimeter may also be divided into 360 equally spaced divisions called *degrees.* Therefore,

$$2\pi \text{ rad} = 360° \tag{A-3}$$

The degree may be further subdivided into 60 divisions called *minutes,* and the minutes may be subdivided into 60 divisions called *seconds.* This is a fractional system of measuring angles that dates back to the Babylonians.

A.2   SOLID ANGLE

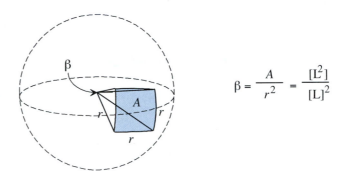

$$\beta = \frac{A}{r^2} = \frac{[L^2]}{[L]^2}$$

TABLE A.2
Solid angle conversion factors (Reference)

	Square Degree	Square Minute	Square Second	Steradian	Sphere
1 square degree =	1	$(60)^2$	$(3600)^2$	$(\pi/180)^2$	$(\pi/4)(180)^{-2}$
1 square minute =	$(1/60)^2$	1	$(60)^2$	$(\pi/10{,}800)^2$	$(\pi/4)(10{,}800)^{-2}$
1 square second =	$(1/3600)^2$	$(1/60)^2$	1	$(\pi/648{,}000)^2$	$(\pi/4)(648{,}000)^{-2}$
1 steradian =	$(180/\pi)^2$	$(10{,}800/\pi)^2$	$(648{,}000/\pi)^2$	1	$(4\pi)^{-1}$
1 sphere =	$(4/\pi)(180)^2$	$(4/\pi)(10{,}800)^2$	$(4/\pi)(648{,}000)^2$	4π	1

1 sphere = 2 hemisphere   1 sphere = 8 spherical right angles

A solid angle β is defined as the surface area on the sphere A divided by the radius r squared:

$$\beta \equiv \frac{A}{r^2} \tag{A-4}$$

The surface can be defined by projecting four radii from the center of a sphere and connecting the ends of adjacent radii with circumference segments. If the angle between adjacent radii is one radian, then a square is defined on the sphere surface that has circumference segments of length r. This solid angle is a *steradian,* given by the formula

$$\beta = \frac{A}{r^2} = \frac{r^2}{r^2} = 1 \text{ steradian} \tag{A-5}$$

If the angle between adjacent radii is 1 degree, then the solid angle is a *square degree*; if the angle between adjacent radii is 1 minute, then the solid angle is a *square minute*; and if the angle between adjacent radii is 1 second, then the solid angle is a *square second*. Table A.2 shows the relationship between these various solid angle measurements.

If a sphere is divided into two parts, then the solid angle is a *hemisphere*. If a hemisphere is divided into four equal parts, the solid angle formed is a *spherical right angle*.

A.3   LENGTH

$$d = [L]$$

TABLE A.3
Length conversion factors (Reference)

	cm	m	km	in	ft	mi [e]
1 centimeter =	1	0.01	1.0000 E–05	0.3937	0.03281	6.214 E–06
1 meter =	100	1	0.001	39.37	3.281	6.214 E–04
1 kilometer =	1.00 E+05	1000	1	3.937 E+04	3281	0.6214
1 inch [b] =	2.54000	0.02540	2.540 E–05	1	0.08333	1.578 E–05
1 foot [a] =	30.48000	0.304800	3.048 E–04	12	1	1.894 E–04
1 U.S. statute mile =	1.609 E+05	1609	1.609	6.336 E+04	5280	1

1 nautical mile (n. mile) [f]=1852 m=1.151 mi=6076 ft	1 rod (rd) = 1 pole = 1 perch = 16.5 ft	1 fermi (fm) [j] = 1.00 E–15 m
1 ångström (Å) [k] = 1.00 E–10 m	1 yard (yd) = 3 ft	1 micron (μ) [l] = 1.00 E–06 m
1 light-year (ly) [g] = 9.4606 E+12 km	1 bolt of cloth = 120 ft	1 printer's pica = 0.16604 in
1 parsec (pc) [h] = 3.086 E+13 km	1 mil [d] = 1 thou = 0.001 in	1 printer's pica = 12 points
1 astronomical unit (i) = 1.496 E+08 km	1 pace = 30 in	1 fathom (fath) [c] = 6 ft
1 statute league = 2640 fathoms	1 cable [m] = 120 fathoms	1 cubit = 18 in
1 chain (ch) = 66 ft = 100 Gunter's links (li)	1 palm = 3 in	1 span = 9 in
1 furlong (fur) = 660 ft = 1/8 mi	1 hand = 4 in	1 skein = 360 ft

[a] The *foot* has been used in England for over 1000 years and is approximately equal to the length of a man's foot.
[b] The *inch* is derived from "ynce," the Anglo-Saxon word for twelfth part.
[c] A *fathom* is used to describe the depth of the sea. It is approximately the distance between the hands when the arms are out-stretched; its name is derived from the Anglo-Saxon word for "embrace."
[d] The *mil* is equal to one thousandth of an inch and is not to be confused with the millimeter. It is commonly used in metal machining.
[e] The *mile* traces to the Romans and is about equal to 1000 double paces (about 5 ft).
[f] The *nautical mile* is the average meridian length of 1 minute of latitude, a definition that makes navigation easier.
[g] The *light-year* is the distance light travels in 1 year.
[h] The *parsec* is the height of an isosceles triangle of which the base is equal to the diameter of the earth's orbit around the sun, and the angle opposite that base is 1″.
[i] An *astronomical unit* is approximately equal to the mean distance from the earth to the sun.
[j] The *fermi* is used to measure nuclear distances.
[k] The *ångström* is used to measure atomic distances (a hydrogen atom is approximately 1 Å).
[l] The *micron* is slang for "micrometer" and is not SI.
[m] The *cable* is used to measure lengths at sea and dates back to the middle of the 16th century.

A.4   AREA

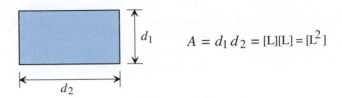

$$A = d_1 d_2 = [L][L] = [L^2]$$

TABLE A.4
Area conversion factors (Reference)

	m²	cm²	ft²	in²
1 square meter =	1	1.000 E+04	10.76	1550
1 square centimeter =	1.000 E–04	1	0.001076	0.1550
1 square foot =	0.09290	929.0	1	144
1 square inch =	6.452 E–04	6.452	0.006944	1

1 square mile = 2.788 E+07 ft² = 640 acres
1 yd² = 9 ft²
1 square rod = 30.25 yd² = 272.25 ft²
1 rood = 40 square rod
1 acre [b] = 4 roods = 160 square rods = 43,560 ft²

1 are (a) [a] = 100 m²
1 hectare (ha) [a] = 100 are = 10,000 m² = 2.471 acres
1 barn (b) [d] = 1.0000 E–28 m²
1 circular mil (cir mils) [c] = (0.001 in)²π/4 = 7.854 E–07 in²
1 U.S. township = 36 mi² = 36 sections

[a] An area 10 m on a side is an *are* and an area 100 m on a side is a *hectare* (i.e., 100 are). Both the are and hectare are used in international agriculture to measure land area.
[b] The *acre,* which has been in existence since about 1300, is the approximate area that a yoke of oxen could plow in a day.
[c] A *circular mil* is the cross-sectional area of a circle that is 1 mil (0.001 in) in diameter. It was first used to measure the cross-sectional area of wire.
[d] The *barn* is used to measure the effective target area of atomic nuclei when bombarded with particles. The unit was invented in 1942 as Manhattan Project code; it probably derives from the expression "I bet you couldn't hit the broadside of a barn."

A.5   VOLUME

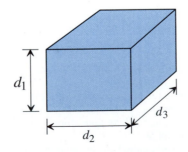

$$V = d_1 d_2 d_3 = [L][L][L] = [L^3]$$

TABLE A.5
Volume conversion factors (Reference)

	m^3	cm^3	L	ft^3	in^3
1 cubic meter =	1	1.000 E+06	1000	35.31	6.102 E+04
1 cubic centimeter =	1.000 E–06	1	0.001	3.531 E–05	0.06102
1 liter =	0.001000	1000	1	0.03531	61.02
1 cubic foot =	0.02832	2.832 E+04	28.32	1	1728
1 cubic inch =	1.639 E–05	16.39	1.639 E–02	5.787 E–04	1

1 acre-foot [d] = 43,560 ft^3
1 board-foot (fbm or bd-ft) [e] = 144 in^3
1 barrel (bbl) [h] = 42 gal
1 U.K. gallon = 1 Imperial gallon = 1.2009 U.S gallon [c]

1 stere (st) [a] = 1 m^3
1 yd^3 = 27 ft^3
1 masonry perch = 24.75 ft^3

1 λ = 1 μL [b]
1 cord (wood) [f] = 128 ft^3
1 cord-foot [g] = 16 ft^3
1 ft^3 = 7.4805195 U.S. gallon (liq)

[a] The *stere* is no longer recommended.
[b] 1 μL is sometimes called 1 λ, but the use of this unit is not recommended.
[c] The *gallon* was first mentioned in 1342 and was given legal status in 1602. The U.S. gallon originated with the old English wine gallon in Colonial times. The *Imperial gallon* (which is about 20% larger than the U.S. gallon) is defined by a 1963 British law as the volume occupied by 10 lb_m of distilled water provided the water has a density of 0.998859 g/mL weighed in air with a density of 0.001217 g/mL against weights with a density of 8.136 g/mL.
[d] An *acre-foot* is the volume when one acre is covered by water with 1 ft depth. This unit is commonly used in agricultural irrigation.
[e] A *board-foot* corresponds to the volume occupied by a board that is 1 ft × 1 ft × 1 in.
[f] The *cord* describes the volume of a wood stack that measures 4 ft × 4 ft × 8 ft.
[g] A *cord-foot* describes the volume of a wood stack that measures 4 ft × 4 ft × 1 ft.
[h] U.S. petroleum barrel.

TABLE A.6
Customary units of volume (Reference)

United Kingdom (Liquids and Solids)

20 minims (min)	= 1 scruple	= 1.1838 E–06 m^3
3 scruples	= 1 fluid drachm	= 3.5515 E–06 m^3
8 fluid drachms	= 1 fluid ounce (fl oz)	= 2.8413 E–05 m^3
5 fluid ounces	= 1 gill or noggin	= 1.4207 E–04 m^3
4 gills	= 1 pint (pt)	= 5.6825 E–04 m^3
2 pints	= 1 quart (qt)	= 1.1365 E–03 m^3
2 quarts	= 1 pottle or quartern (dry)	= 2.2730 E–03 m^3
2 quarterns (dry)	= 1 gallon (gal)	= 4.5461 E–03 m^3
2 gallons	= 1 peck (pk)	= 9.0919 E–03 m^3
4 pecks	= 1 bushel (bu)	= 3.6368 E–02 m^3
9 gallons	= 1 firkin	= 4.0914 E–02 m^3
9 pecks	= 1 kilderkins	= 8.1830 E–02 m^3
3 bushels	= 1 sack or bag	= 1.0910 E–01 m^3
36 gallons	= 1 barrel (bbl)	= 1.6365 E–01 m^3
8 bushels	= 1 quarter or seam	= 2.9094 E–01 m^3
640 gallons	= 1 lasts	= 2.9094 m^3

United States (Liquid)

60 minims (min)	= 1 fluid dram (fl dr)	= 3.6967 E–06 m^3
3 teaspoons (t or tsp)	= 1 tablespoon (T or Tbsp)	= 1.4787 E–05 m^3
2 tablespoons	= 1 fluid ounce (fl oz)	= 2.9574 E–05 m^3
8 fluid drams	= 1 fluid ounce (fl oz)	= 2.9574 E–05 m^3
4 fluid ounces	= 1 gill	= 1.1829 E–04 m^3
2 gills	= 1 cup	= 2.3659 E–04 m^3
2 cups	= 1 liquid pint (pt)	= 4.7318 E–04 m^3
2 liquid pints	= 1 liquid quart (qt)	= 9.4635 E–04 m^3
4 liquid quarts	= 1 gallon (gal)	= 3.7854 E–03 m^3
9 gallons	= 1 firkin	= 3.4068 E–02 m^3
31.5 gallons	= 1 barrel (bbl)*	= 1.1924 E–01 m^3
63 gallons	= 1 hogshead (hhd)	= 2.3847 E–01 m^3
84 gallons	= 1 puncheon	= 3.1797 E–01 m^3
126 gallons	= 1 U.K. butt	= 4.7696 E–01 m^3
252 gallons	= 1 tun	= 9.5392 E–01 m^3

United States (Dry)

2 dry pints	= 1 dry quart (qt)	= 1.1012 E–03 m^3
4 dry quarts	= 1 dry gallon (gal)	= 4.4049 E–03 m^3
2 dry gallons	= 1 peck (pk)	= 8.8098 E–03 m^3
4 pecks	= 1 bushel (bu)	= 3.5239 E–02 m^3
105 dry quarts	= 1 dry barrel (bbl)*	= 1.1563 E–01 m^3

*Not to be confused with a U.S. petroleum barrel.

A.6   MASS

 [M]

TABLE A.7
Mass unit conversions (Reference)

	g	kg	lb$_m$	slug
1 gram-mass =	1	0.001	0.002205	6.852 E–05
1 kilogram-mass =	1000	1	2.205	0.06852
1 pound-mass [c] =	453.6	0.4536	1	0.03108
1 slug =	1.4594 E+04	14.594	32.174	1

1 grain [b] = 6.479891 E–05 kg
1 short hundred weight = 100 lb$_m$
1 short ton = 2000 lb$_m$
1 tonne (t) = 1 metric ton = 1000 kg
1 metric carat [d] = 2.000 E–04 kg
1 point = 0.01 metric carat
1 γ [a] = 1 μg = 1.000 E–09 kg

1 glug = 980.665 g = 0.980665 kg
1 mug = 1 metric slug = 1 par =1 TME = 9.80665 kg
1 unified atomic mass unit (u) [e] = 1 dalton = 1.6605402 E–27 kg
1 atomic mass unit, chem. (amu) [e] = 1.66024 E–27 kg
1 atomic mass unit, phys. (amu) [e] = 1.65979 E–27 kg
1 eV of equivalent mass [f] = 1.7827 E–36 kg

[a] The symbol "γ" is used to represent 1 μg, but its use is discouraged.
[b] The *grain* dates back to the 16th century and is thought to be equal to the weight of a wheat grain.
[c] The *pound* originated with the Roman Libra (327 g). The Imperial Standard Pound was defined in 1855 as the mass of platinum with given dimensions. In 1963, the pound was defined as 0.45359237 kg exactly, a number chosen because it is evenly divided by seven to ease the conversion from grains to grams.
[d] Precious stones are measured in *metric carats,* which correspond to 200 mg.
[e] The *atomic mass unit* was originally intended to be the mass of a single hydrogen atom, the lightest element. In 1885, it was suggested that more elements would have integer numbers for their atomic weights if the atomic mass unit were defined using 1/16 the mass of oxygen. Chemists used oxygen in its natural abundance (2480:5:1 ^{16}O:^{18}O:^{17}O) whereas physicists used isotopically pure ^{16}O for their standard. Thus, there was a slight discrepancy between the scales used by chemists and physicists (272 parts per million). It was later found that expressing the atomic mass unit as 1/12th the mass of a single carbon-12 atom allowed even more elements to have masses that were integer numbers. Thus, the *unified atomic mass unit* was established, which had the added benefit of eliminating the discrepancy between the chemist and physicist scales.
[f] The famous Einstein relationship $E=mc^2$ showed that when mass is destroyed, energy is produced (and vice versa). The amount of energy E is found by multiplying the destroyed mass m by the speed of light c squared. Thus, physicists and nuclear engineers sometimes express mass in energy units, such as electron volts (eV).

TABLE A.8
Customary units of mass (Reference)

Avoirdupois Weights		Apothecaries Weights [b]		Troy Weights [c]	
1 pound (lb avdp) [a]		**1 pound (lb ap)**		**1 pound (lb t)**	
= 7000 grains [d]		**= 5760 grains [d]**		**= 5760 grains [d]**	
16 drams (dr avdp)	= 1 ounce (oz)	20 grains	= 1 scruple (s ap)	24 grains	= penny weight (dwt)
16 ounces	= 1 pound (lb avdp)	3 scruples	= 1 U.K. drachm (dr ap)	20 penny weights	= 1 ounce (oz t)
14 pounds	= 1 stone	3 scruples	= 1 U.S. dram (dr ap)	12 ounce (oz t)	= 1 pound (lb t)
28 pounds	= 1 quarter	8 drachm or dram	= 1 ounce (oz ap)		
112 pounds	= 1 long hundred weight (cwt)	12 ounce (oz ap)	= 1 pound (lb ap)		
252 pounds	= 1 wey				
2240 pounds	= 1 long ton				

[a] The common pound with which we are familiar (and the pound indicated by the symbol "lb_m") is the avoirdupois pound.
[b] The apothecary scale is not used anymore.
[c] The troy scale is used in the United States for weighing precious metals.
[d] The grain is the same in all systems.

A.7   DENSITY

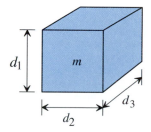

$$\rho = \frac{m}{d_1\,d_2\,d_3} = \frac{[M]}{[L][L][L]} = \frac{[M]}{[L^3]}$$

TABLE A.9
Density conversion factors (Reference)

	g/cm³	kg/m³	lb_m/ft³	lb_m/in³	slug/ft³
1 gram per cubic centimeter =	1	1000	62.43	0.03613	1.940
1 kilogram per cubic meter =	0.001	1	0.06243	3.613 E–05	0.001940
1 pound-mass per cubic foot =	0.01602	16.02	1	5.787 E–04	0.03108
1 pound-mass per cubic inch =	27.68	2.768 E+04	1728	1	53.71
1 slug per cubic foot =	0.5154	515.4	32.174	0.01862	1

Density can also be expressed by *specific gravity* SG, a dimensionless number formed by dividing the density of substance A ρ_A by the density of a reference substance ρ_R:

$$SG = \frac{\rho_A}{\rho_R} \tag{A-6}$$

Although any reference may be used, the most common reference substance is water at its maximum density (4°C, 1.000 g/cm³).

A.8   TIME

TABLE A.10
Time conversion factors (Reference)

	yr	d	h	min	s
1 year [a] =	1	365.24	8.766 E+03	5.259 E+05	3.1557 E+07
1 day [c] =	2.738 E–03	1	24	1440	8.640 E+04
1 hour [d] =	1.141 E–04	4.167 E–02	1	60	3600
1 minute [e] =	1.901 E–06	6.944 E–04	1.667 E–02	1	60
1 second [e] =	3.169 E–08	1.157 E–05	2.778 E–04	1.667 E–02	1

1 year = 365.24 solar days [c]	1 year = 366.24 sidereal days [b]	1 week = 7 days
1 mean solar day [c] = 86,400 s	1 sidereal day [b] = 86,164 s	1 fortnight = 2 weeks

[a] A *year* is the time required for the earth to return to a given position as it orbits the sun. Our calendar is adjusted to the *tropical year*, the time it takes for the earth to orbit the sun between successive vernal equinoxes (March 21, the spring date in which light and dark are equal).

[b] A *sidereal day* is the mean time taken for the earth to complete one revolution as determined by comparing the earth's position to distant stars.

[c] A *solar day* is the mean time required for the sun to return to a fixed position (e.g., overhead) in the sky. The solar day and sidereal day differ. The solar day is slightly longer because the sun is viewed from a different position as the earth orbits the sun. In common parlance, we refer to a solar day, not a sidereal day. It has been known since the Egyptians and Babylonians that there are $365\frac{1}{4}$ solar days per year.

[d] In ancient times, the day was divided into 24 time fractions which we call *hours*. Light and darkness were each divided into 12 equal time fractions regardless of the time of year. According to the season, the length of the dark-hour differed from the light-hour. When mechanical clocks were invented, the length of the hour was standardized. In England, each community kept its own local time; each community was completely independent of the others. In 1880, Greenwich mean time was established as the official time throughout England. Today, most of the world has agreed to standardize on Greenwich mean time.

[e] The *minute* and *second* of time trace to the Babylonians, who used units of 60. Efforts to decimalize time have proved unsuccessful.

A.9   SPEED/VELOCITY

$$\text{Speed, Velocity} = \frac{d}{t_2 - t_1} = \frac{[L]}{[T]} = [L/T]$$

TABLE A.11
Speed/velocity conversion factors (Reference)

	m/s	cm/s	ft/s	km/h	mi/h (mph)	knot
1 meter per second =	1	100	3.281	3.6	2.237	1.944
1 centimeter per second =	0.01	1	0.03281	0.036	0.02237	0.01944
1 foot per second =	0.3048	30.48	1	1.097	0.6818	0.5925
1 kilometer per hour =	0.2778	27.78	0.9113	1	0.6214	0.5400
1 mile per hour =	0.4470	44.70	1.467	1.609	1	0.8690
1 nautical mile per hour =	0.5144	51.44	1.688	1.852	1.151	1

1 knot = 1 nautical mile per hour   1 mi/min = 88.00 ft/s = 60.00 mi/h

A.10   FORCE

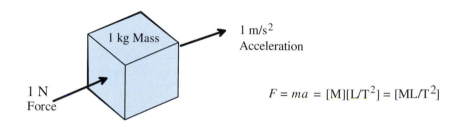

1 kg Mass

1 m/s^2
Acceleration

1 N
Force

$$F = ma = [M][L/T^2] = [ML/T^2]$$

TABLE A.12
Force conversion factors (Reference)

	N	dyne	pdl	kg$_f$	g$_f$	lb$_f$
1 newton =	1	1.00 E+05	7.233	0.1020	102.0	0.2248
1 dyne =	1.00 E-05	1	7.233 E-05	1.020 E-06	0.001020	2.248 E-06
1 poundal =	0.1383	1.383 E+04	1	0.01410	14.10	0.03108
1 kilogram-force =	9.807	9.807 E+05	70.93	1	1000	2.205
1 gram-force =	0.009807	980.7	0.07093	0.001	1	0.002205
1 pound-force =	4.448	4.448 E+05	32.174	0.4536	453.6	1

1 pound-force = 16 ounce-force   1 kilopond [b] = 1 kg$_f$ 1 kip [a] = 1000 lb$_f$
1 ton-force = 2000 lb$_f$ 1 fors [c] = 1 g$_f$

[a] The *kip* (for *Ki*lo *I*mperial *P*ound) is sometimes used to describe the load on a structure.
[b] The *kilopond* is used in Germany for "kilogram-force."
[c] The *fors* (Latin for "force") was proposed in 1956 as an alternate name for "gram-force."

A.11   PRESSURE

$$P = \frac{F}{d_1 d_2} = \frac{[ML/T^2]}{[L][L]} = [M/LT^2]$$

TABLE A.13
Pressure conversion factors (Reference)

	Pa	dyne/cm^2	lb$_f$/ft^2	lb$_f$/in^2 (psi)	atm	cm-Hg	in-H$_2$O
1 newton per square meter =	1	10	0.02089	1.450 E–04	9.869 E–06	7.501 E–04	0.004015
1 dyne per square centimeter =	0.1	1	0.002089	1.450 E–05	9.869 E–07	7.501 E–05	4.015 E–04
1 pound-force per square foot =	47.88	478.8	1	0.006944	4.725 E–04	0.03591	0.1922
1 pound-force per square inch =	6895	6.895 E+04	144	1	0.06805	5.171	27.68
1 standard atmosphere [c] =	1.013 E+05	1.013 E+06	2116	14.696	1	76	406.8
1 centimeter [d] of mercury at 0°C =	1333	1.333 E+04	27.84	0.1934	0.01316	1	5.353
1 inch [d] of water at 4°C =	249.1	2491	5.202	0.03613	0.002458	0.1868	1

1 kg$_f$/m^2 = 9.806650 Pa
1 atm = 2.493 ft-Hg = 33.90 ft-H$_2$O = 27,714 ft-air (1 atm, 60°F) [d]
1 bar [a] = 1 barye = 1.00 E+06 dyne/cm^2 = 0.1 MPa = 100 kPa ≈ 1 atm
1 millibar (mb) = 1.00 E+03 dyne/cm^2 = 1000 microbar (μb) [b]
1 torr [e] = (101325/760) Pa ≈ 1 μm-Hg = 0.1 cm-Hg

1 kip/in^2 (ksi) = 1000 lb$_f$/in^2
1 technical atmosphere [c] = 1 kg$_f$/cm^2
1 micron pressure = 1 μm-Hg [d]
1 g$_f$/cm^2 = 980.665 dyne/cm^2

[a] The bar is most commonly employed in meteorology because it is approximately equal to the atmospheric pressure on earth. Although the bar is not properly SI, its use is temporarily tolerated because it is so widespread. The *barye* was the original name given to this unit of pressure in 1900, but it has been shortened to "bar."
[b] Although there is no proper abbreviation for the bar, the *millibar* (mb) and *microbar* (μb) abbreviations are sometimes used.
[c] Because the atmospheric pressure changes (in fact, meteorologists measure it to predict weather changes), a *standard atmosphere* $P°$ has been defined as 101,325.0 Pa. The *technical atmosphere* is defined as 1 kg$_f$/cm^2. Unless otherwise specified, an "atmosphere" is generally the "standard atmosphere." The use of *atmosphere* for pressure measurements is discouraged by SI, but its use will probably continue because it is easily visualized.
[d] The simplest way to measure pressure is with a *manometer*, a U-shaped tube filled with liquid. Differences in pressure acting on each liquid column change the liquid levels, which are then easily read using a meterstick. For accurate work, the conversion factors in Table A.13 may be used only if the temperature of the liquid is controlled [4°C for water, 0°C for mercury (Hg)]. (Alternatively, tables listing the liquid density as a function of temperature may be used to correct the reading, provided the manometer temperature is known.) Also, the local acceleration due to gravity (g) affects the reading. The values given in Table A.13 use the standard acceleration due to gravity ($g°$).
[e] The *torr* differs from a mm-Hg by less than 1 part in 7 million. The use of *torr* is discouraged by SI.

A.12   ENERGY

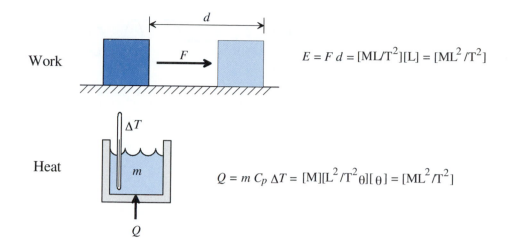

Work

$$E = F\,d = [ML/T^2][L] = [ML^2/T^2]$$

Heat

$$Q = m\,C_p\,\Delta T = [M][L^2/T^2\theta][\theta] = [ML^2/T^2]$$

TABLE A.14
Energy conversion factors (Reference)

	J	erg	ft·lb$_f$	cal	Btu	kW·h	hp·h
1 joule =	1	1.000 E+07	0.7376	0.2390	9.485 E–04	2.778 E–07	3.725 E–07
1 erg =	1.000 E–07	1	7.376 E–08	2.390 E–08	9.485 E–11	2.778 E–14	3.725 E–14
1 foot pound$_f$ =	1.356	1.356 E+07	1	0.3240	0.001286	3.766 E–07	5.051 E–07
1 calorie [b] =	4.184	4.184 E+07	3.086	1	0.003968	1.162 E–06	1.559 E–06
1 Brit. thermal unit [c] =	1054	1.054 E+10	777.6	252.0	1	2.929 E–04	3.928 E–04
1 kilowatt hour [d] =	3.600 E+06	3.600 E+13	2.655 E+06	8.606 E+05	3414	1	1.341
1 horsepower hour [d] =	2.685 E+06	2.685 E+13	1.980 E+06	6.414 E+05	2545	0.7457	1

1 electron volt (eV) [f] = 1.60217733 E–19 J
1 kg$_f$·m = 9.806650 J
1 V·C = 1 J
1 (dyne/cm^2)·cm^3 = 1 erg [e]
1 atm·ft^3 = 2116 ft·lb$_f$ [e]
1 ton (nuclear equivalent TNT) = 4.184 E+09 J

1 kcal [a] = 1 calorie (kg)
1 g$_f$·cm = 980.6650 erg
1 V·A·s = 1 J
1 atm·L = 101.3 J [e]
1 psia·ft^3 = 144 ft·lb$_f$ [e]

1 W·h = 3600 J [d]
1 W·s = 1 J [d]
1 Pa·m^3 = 1 J [e]
1 atm·cm^3 = 0.1013 J [e]
1 bar·cm^3 = 0.1 J [e]

[a] *Kilocalorie* is the heat required to raise 1 kg water by 1 K. Because the heat capacity of water is not constant, a variety of kilocalories are defined. This is the *thermochemical* kilocalorie, the most commonly used.
[b] *Calorie* is the heat required to raise 1 g water by 1 K. In diet books, the energy content in food is usually expressed in *calories*, but actually *kilocalories* are meant. Sometimes dietitians use *Calorie* to mean *kilocalorie*. Because the heat capacity of water is not constant, a variety of calories are defined. This is the *thermochemical* calorie, the most commonly used.
[c] *British thermal unit* is the heat required to raise 1 lb$_m$ water by 1 F°. Because the heat capacity of water is not constant, a variety of Btus are defined. This is the *thermochemical* Btu, the most commonly used.
[d] Energy = power × time. These units can be visualized as answering the question "how much energy is expended if a 1-kW (1-hp) motor operates for 1 hour?"
[e] Energy = pressure × volume
[f] The energy required to move a single electron through a vacuum with 1 volt of potential is an *electron volt*.

A.13 POWER

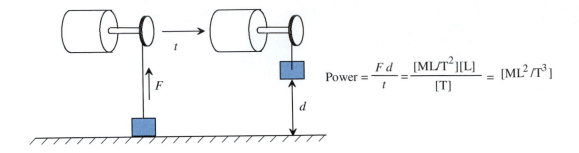

$$\text{Power} = \frac{F\,d}{t} = \frac{[ML/T^2][L]}{[T]} = [ML^2/T^3]$$

TABLE A.15
Power conversion factors (Reference)

	W	kW	ft·lb$_f$/s	hp	cal/s	Btu/h
1 watt [a] =	1	0.001	0.7376	0.001341	0.2390	3.414
1 kilowatt =	1000	1	737.6	1.341	239.0	3414
1 foot-pound$_f$ per second =	1.356	0.001356	1	0.001818	0.3240	4.629
1 horsepower [b] =	745.7	0.7457	550	1	178.2	2546
1 calorie per second =	4.184	0.004184	3.086	0.005611	1	14.29
1 British thermal unit per hour =	0.2929	2.929 E–04	0.2160	3.928 E–04	0.07000	1

1 W = 1.00 E+07 erg/s   1 ft·lb$_f$/s = 60 ft·lb$_f$/min = 3600 ft·lb$_f$/h   1 hp = 33,000 ft·lb$_f$/min = 550 ft·lb$_f$/s
1 hp (electric) ≡ 746 W   1 ton of refrigeration [d] = 12,000 Btu/h
1 hp = 0.0760181 hp (boiler) = 0.999598 hp (electric) = 1.01387 hp (metric) = 0.999540 hp (water) [c]

[a] A *watt* is a J/s.
[b] In 1782, James Watt (1736–1819) devised the *horsepower* to help him sell steam engines. He assumed a horse could pull with a force of 180 lb$_f$ and, when harnessed to a capstan, would walk a 24-ft diameter circle $2\frac{1}{2}$ times each minute. This was a work expenditure of 32,400 ft·lb$_f$/min, which he rounded to 33,000 ft·lb$_f$/min (550 ft·lb$_f$/s). Engines are often rated in *brake horsepower* (bhp) or *shaft horsepower*, which is the power available at the turning drive shaft.
[c] A *boiler horsepower* (bhp) is the amount of heat needed to evaporate 34.5 lb$_m$/h of water at 212°F. A *metric horsepower* is the power required to raise 75 kg 1 meter per second.
[d] A *ton of refrigeration* is a U.S. term that describes the amount of refrigeration required to freeze 1 ton (2000 lb$_m$) per day of water at 32°F.

A.14 AMOUNT OF SUBSTANCE

The *mole* is the number of atoms in 0.012 kg (12 g) of carbon-12. This number is given special recognition as *Avogadro's constant* N_A, which is

$$N_A = 6.0221367 \times 10^{23} \text{ atoms/mol} \tag{A-7}$$

The *coulomb* C is the number of electrons that flow in a 1-ampere current in 1 second. The number of electrons in a coulomb N_C is

$$N_C = 6.24150636 \times 10^{18} \text{ electrons/C} \tag{A-8}$$

The ratio of Avogadro's number to the coulomb is called the *Faraday constant F*

$$F = \frac{N_A}{N_C} = 96{,}485.309 \text{ C/mol} \tag{A-9}$$

The *mol* is sometimes called the *gram-mole*. The *kilogram-mole* (kmol) is the number of atoms in 12 kg of carbon-12, the *pound-mole* (lbmol) is the number of atoms in 12 lb$_m$ of carbon-12, and the *ton-mole* is the number of atoms in 12 tons of carbon-12. The number of atoms in each of these units is calculated as:

$$\frac{6.022 \times 10^{23} \text{ atoms}}{\text{mol}} \times \frac{1000 \text{ mol}}{\text{kmol}} = 6.022 \times 10^{26} \frac{\text{atoms}}{\text{kmol}}$$

$$\frac{6.022 \times 10^{23} \text{ atoms}}{\text{mol}} \times \frac{453.6 \text{ mol}}{\text{lbmol}} = 2.732 \times 10^{26} \frac{\text{atoms}}{\text{lbmol}}$$

$$\frac{6.022 \times 10^{23} \text{ atoms}}{\text{mol}} \times \frac{453.6 \text{ mol}}{\text{lbmol}} \times \frac{2000 \text{ lbmol}}{\text{ton-mole}} = 5.463 \times 10^{29} \frac{\text{atoms}}{\text{ton-mol}}$$

Further Readings

Jerrard, H. G., and D. B. McNeill. *A Dictionary of Scientific Units,* 6th ed. rev. Netherlands: Kluwer Academic Publishers, 1992.

Klein, H. A. *The World of Measurements.* New York: Simon and Schuster, 1974.

National Institute of Standards and Technology. *The International System of Units (SI).* NIST Special Publication 330, U.S. Department of Commerce, 1991.

Weast, R. C. *CRC Handbook of Chemistry and Physics,* 87th ed. West Palm Beach: CRC Press, 2006 (Also downloadable through crcpress.com).

APPENDIX B

NATIONAL SOCIETY OF PROFESSIONAL ENGINEERS
Code of Ethics for Engineers

Preamble

Engineering is an important and learned profession. The members of the profession recognize that their work has a direct and vital impact on the quality of life for all people. Accordingly, the services provided by engineers require honesty, impartiality, fairness and equity, and must be dedicated to the protection of the public health, safety and welfare. In the practice of their profession, engineers must perform under a standard of professional behavior which requires adherence to the highest principles of ethical conduct on behalf of the public, clients, employers and the profession.

I. Fundamental Canons

Engineers, in the fulfillment of their professional duties, shall:

1. Hold paramount the safety, health and welfare of the public in the performance of their professional duties.
2. Perform services only in areas of their competence.
3. Issue public statements only in an objective and truthful manner.
4. Act in professional matters for each employer or client as faithful agents or trustees.
5. Avoid deceptive acts in the solicitation of professional employment.

II. Rules of Practice

1. Engineers shall hold paramount the safety, health and welfare of the public in the performance of their professional duties.
   a. Engineers shall at all times recognize that their primary obligation is to protect the safety, health, property and welfare of the public. If their professional judgement is overruled under circumstances where the safety, health, property or welfare of the public are endangered, they shall notify their employer or client and such other authorities as may be appropriate.
   b. Engineers shall approve only those engineering documents which are safe for public health, property and welfare in conformity with accepted standards.

decisions with respect to professional services solicited or provided by them or their organizations in private or public engineering practice.

    e. Engineers shall not solicit or accept a professional contract from a government body on which a principal or officer of their organization serves as a member.

5. Engineers shall avoid deceptive acts in the solicitation of professional employment.

    a. Engineers shall not falsify or permit misrepresentation of their, or their associates', academic or professional qualifications. They shall not misrepresent or exaggerate their degree of responsibility in or for the subject matter of prior assignments. Brochures or other presentations incident to the solicitation of employment shall not misrepresent pertinent facts concerning employers, employees, associates, joint ventures or past accomplishments with the intent and purpose of enhancing their qualifications and their work.

    b. Engineers shall not offer, give, solicit or receive, either directly or indirectly, any political contribution in an amount intended to influence the award of a contract by public authority, or which may be reasonably construed by the public of having the effect or intent to influence the award of a contract. They shall not offer any gift, or other valuable consideration in order to secure work. They shall not pay a commission, percentage or brokerage fee in order to secure work except to a bona fide employee or bona fide established commercial or marketing agencies retained by them.

III. Professional Obligations

1. Engineers shall be guided in all their professional relations by the highest standards of integrity.

    a. Engineers shall admit and accept their own errors when proven wrong and refrain from distorting or altering the facts in an attempt to justify their decisions.

    b. Engineers shall advise their clients or employers when they believe a project will not be successful.

    c. Engineers shall not accept outside employment to the detriment of their regular work or interest. Before accepting any outside employment they will notify their employers.

    d. Engineers shall not attempt to attract an engineer from another employer by false or misleading pretenses.

    e. Engineers shall not actively participate in strikes, picket lines, or other collective coercive action.

    f. Engineers shall avoid any act tending to promote their own interest at the expense of the dignity and integrity of the profession.

2. Engineers shall at all times strive to serve the public interest.

    a. Engineers shall seek opportunities to be of constructive service in civic affairs and work for the advancement of the safety, health and well-being of their community.

    b. Engineers shall not complete, sign or seal plans and/or specifications that are not of a design safe to the public health and welfare and in conformity with accepted engineering standards. If the client or employer insists on such

    c. Engineers shall not reveal facts, data or information ob
       sional capacity without the prior consent of the client or e
       authorized or required by law or this Code.

    d. Engineers shall not permit the use of their name or firm name
       business ventures with any person or firm which they have r
       is engaging in fraudulent or dishonest business or professiona

    e. Engineers having knowledge of any alleged violation of th
       cooperate with the proper authorities in furnishing such informa
       tance as may be required.

2. Engineers shall perform services only in the areas of their competen

    a. Engineers shall undertake assignments only when qualified by ed
       experience in the specific technical fields involved.

    b. Engineers shall not affix their signatures to any plans or documents
       with subject matter in which they lack competence, nor to any plan o
       ment not prepared under their direction and control.

    c. Engineers may accept assignments and assume responsibility for coo
       tion of an entire project and sign and seal the engineering documents fo
       entire project, provided that each technical segment is signed and sealed
       by the qualified engineers who prepared the segment.

3. Engineers shall issue public statements only in an objective and truthful mann

    a. Engineers shall be objective and truthful in professional reports, statemen
       or testimony. They shall include all relevant and pertinent information i
       such reports, statements and testimony.

    b. Engineers may express publicly a professional opinion on technical subjects
       only when that opinion is founded upon adequate knowledge of the facts and
       competence in the subject matter.

    c. Engineers shall issue no statements, criticisms or arguments on technical
       matters which are inspired or paid for by interested parties, unless they have
       prefaced their comments by explicitly identifying the interested parties on
       whose behalf they are speaking, and by revealing the existence of any inter-
       est the engineers may have in the matters.

4. Engineers shall act in professional matters for each employer or client as faithful agents or trustees.

    a. Engineers shall disclose all known or potential conflicts of interest to their
       employers or clients by promptly informing them of any business associa-
       tion, interest, or other circumstances which could influence or appear to
       influence their judgment or the quality of their services.

    b. Engineers shall not accept compensation, financial or otherwise, from more
       than one party for services on the same project, or for services pertaining to
       the same project, unless the circumstances are fully disclosed to, and agreed
       to by, all interested parties.

    c. Engineers shall not solicit or accept financial or other valuable consideration
       directly or indirectly, from contractors, their agents, or other parties in connec-
       tion with work for employers or clients for which they are responsible.

    d. Engineers in public service as members, advisors or employees of a govern-
       mental or quasi-governmental body or department shall not participate in

unprofessional conduct, they shall notify the proper authorities and withdraw from further service on the project.

 c. Engineers shall endeavor to extend public knowledge and appreciation of engineering and its achievements and to protect the engineering profession from misrepresentation and misunderstanding.

3. Engineers shall avoid all conduct or practice which is likely to discredit the profession or deceive the public.

 a. Engineers shall avoid the use of statements containing a material misrepresentation of fact or omitting a material fact necessary to keep statements from being misleading or intended or likely to create an unjustified expectation, or statements containing prediction of future success.

 b. Consistent with the foregoing, Engineers may advertise for recruitment of personnel.

 c. Consistent with the foregoing, Engineers may prepare articles for the lay or technical press, but such articles shall not imply credit to the author for work performed by others.

4. Engineers shall not disclose confidential information concerning the business affairs or technical processes of any present of former client or employer without his consent.

 a. Engineers in the employ of others shall not without the consent of all interested parties enter promotional efforts or negotiations for work or make arrangements for other employment as a principal or to practice in connection with a specific project for which the Engineer has gained particular and specialized knowledge.

 b. Engineers shall not, without the consent of all interested parties, participate in or represent an adversary interest in connection with a specific project or proceeding in which the Engineer has gained particular specialized knowledge on behalf of a former client or employer.

5. Engineers shall not be influenced in their professional duties by conflicting interests.

 a. Engineers shall not accept financial or other considerations, including free engineering designs, from material or equipment suppliers for specifying their product.

 b. Engineers shall not accept commissions or allowances, directly or indirectly, from contractors or other parties dealing with clients or employers of the Engineer in connection with work for which the Engineer is responsible.

6. Engineers shall uphold the principle of appropriate and adequate compensation for those engaged in engineering work.

 a. Engineers shall not accept remuneration from either an employee or employment agency for giving employment.

 b. Engineers, when employing other engineers, shall offer a salary according to professional qualifications.

7. Engineers shall not attempt to obtain employment or advancement or professional engagements by untruthfully criticizing other engineers, or by other improper or questionable methods.

    a. Engineers shall not request, propose, or accept a professional commission on a contingent basis under circumstances in which their professional judgment may be compromised.

    b. Engineers in salaried positions shall accept part-time engineering work only to the extent consistent with policies of the employer and in accordance with ethical considerations.

    c. Engineers shall not use equipment, supplies, laboratory, or office facilities of an employer to carry on outside private practice without consent.

8. Engineers shall not attempt to injure, maliciously or falsely, directly or indirectly, the professional reputation, prospects, practice or employment of other engineers, nor untruthfully criticize other engineers' work. Engineers who believe others are guilty of unethical or illegal practice shall present such information to the proper authority for action.

    a. Engineers in private practice shall not review the work of another engineer for the same client, except with the knowledge of such engineer, or unless the connection of such engineer with the work has been terminated.

    b. Engineers in governmental, industrial or educational employ are entitled to review and evaluate the work of other engineers when so required by their employment duties.

    c. Engineers in sales or industrial employ are entitled to make engineering comparisons of represented products with products of other suppliers.

9. Engineers shall accept personal responsibility for their professional activities; provided, however, that Engineers may seek indemnification for professional services arising out of their practice for other than gross negligence, where the Engineer's interests cannot otherwise be protected.

    a. Engineers shall conform with state registration laws in the practice of engineering.

    b. Engineers shall not use association with a nonengineer, a corporation, or partnership as a "cloak" for unethical acts, but must accept personal responsibility for all professional acts.

10. Engineers shall give credit for engineering work to those to whom credit is due, and will recognize the proprietary interests of others.

    a. Engineers shall, whenever possible, name the person or persons who may be individually responsible for designs, inventions, writings, or other accomplishments.

    b. Engineers using designs supplied by a client recognize that the designs remain the property of the client and may not be duplicated by the Engineer for others without express permission.

    c. Engineers, before undertaking work for others in connection with which the Engineer may make improvements, plans, designs, inventions, or other records which may justify copyrights or patents, should enter into a positive agreement regarding ownership.

    d. Engineers' designs, data, records, and notes referring exclusively to an employer's work are the employer's property.

11. Engineers shall cooperate in extending the effectiveness of the profession by interchanging information and experience with other engineers and students, and

will endeavor to provide opportunity for the professional development and advancement of engineers under their supervision.

a. Engineers shall encourage engineering employee's efforts to improve their education.

b. Engineers shall encourage engineering employees to attend and present papers at professional and technical society meetings.

c. Engineers shall urge engineering employees to become registered at the earliest possible date.

d. Engineers shall assign a professional engineer duties of a nature to utilize full training and experience, insofar as possible, and delegate lesser functions to subprofessionals or to technicians.

e. Engineers shall provide a prospective engineering employee with complete information on working conditions and proposed status of employment, and after employment will keep employees informed of any changes.

Note: In regard to the question of application of the Code to corporations vis-a-vis real persons, business form or type should not negate nor influence conformance of individuals to the Code. The Code deals with professional services, which services must be performed by real persons. Real persons in turn establish and implement policies within business structures. The Code is clearly written to apply to the Engineer and it is incumbent on a member of NSPE to endeavor to live up to its provisions. This applies to all pertinent sections of the Code.

Publication date as revised: January 2006.
Publication # 1204

APPENDIX C

SUMMARY OF SOME ENGINEERING MILESTONES

Date	Milestone
B.C.	
6000 to 3000	People built permanent houses, cultivated plants, and domesticated animals. Irrigation systems were constructed; plows and animal yokes were used. Wind- and water-powered mills were used to grind grain. Copper ores were mined and transformed into copper and bronze tools. Mathematics was used. Information was written on papyrus, parchment, or clay tablets.
ca. 3050	Earliest evidence of stone masonry in Egypt.
ca. 2930	First pyramid constructed (214 ft).
ca. 2900	Great Pyramid at Gizeh begun. At 481 ft, it was the largest stone building ever erected by ancient humans.
ca. 2000	The Egyptians built irrigation dams and canals.
ca. 1600	The first engineer's handbook, the Rhind Papyrus, was created.
ca. 1500	The palace of Cnossus in Crete was built. It included the first adequate sanitary drains.
ca. 1100	Military engineering was started under the Assyrian king Tiglathpileser I.
ca. 1000	Phoenicians constructed mines.
	King Solomon's temple was built in Jerusalem.
691	Assyrian aqueduct of Jerwan was constructed.
ca. 600	The Egyptians built a canal connecting the Nile and Red Sea.
	The Etruscans built the first arch bridge.
484	Mining was a major source of Greek taxes.
ca. 450	The Greek Empedocles of Akragas drained swamps to prevent disease.
	The Greek Parthenon was constructed.
ca. 300	Appius Claudius constructed the Appian Road in Rome.
	Appius Claudius constructed the Aqua Appia water supply in Rome.
	Greek lighthouse near Alexandria was constructed. It lasted 16 centuries before falling.
ca. 250	Greece's Archimedes designed military machines and screw pump.
ca. 200	The Great Wall of China was completed.
ca. 150	Greece's Hero designed military machines, derricks, presses, rotary steam turbine, odometer, and the hand-pump fire engine.
142	First stone arch bridge Pons Aemilius was constructed.
140	High-level aqueduct Marcia was constructed.
ca. 15	Rome's Marcus Vitruvius Pollio wrote *De Architectura,* which was used as a standard engineering reference work until the Renaissance. The book described land-leveling devices, water supply, time measurement by sundials and water clocks, hoists, derricks, pulleys, pumps, water organs, military "engines of war" (e.g., catapults), and ethics.

A.D.

ca. 45	The Romans constructed a 3.5-mile tunnel to drain rich agricultural lands.
79	The Roman surveyor Frontinius described the Roman 250-mile aqueduct system, which could deliver an estimated 300 million gallons per day to Rome.
80	Roman Coliseum was constructed.
ca. 200	Cast iron was used in China.
ca. 300	Romans constructed a water-powered flour mill in Arles, France, to replace scarce slave labor.
ca. 1000	The abacus calculating machine was introduced to Europe from the Orient.
ca. 1100	Construction of medieval stone fortresses began. They were obsolete by ca. 1500 when gunpowder and cannons could destroy them.
	Windmills were introduced to mill grain, pump, and grind paint and snuff.
ca. 1150	Spanish papermaking became an industry based upon imported Chinese technology.
	Chimneys first appeared in European buildings.
ca. 1200	Black powder was used in Europe.
	Locks for canals were developed in Italy.
ca. 1230	A notebook by Frenchman Wilars de Honecourt described surveying, stone cutting, water-powered saws, and a perpetual motion machine.
ca. 1300	Construction of great Gothic cathedrals began in Europe.
	Spinning wheels were developed to twist fibers to make thread.
	Cast iron was used in Europe.
ca. 1400	Water mills were widely available in European villages.
ca. 1450	Germany's Johann Gutenberg published the first book using a combination of previously known techniques.
ca. 1500	First engineering book, Valturius' *De re militari*, was published.
	Current-driven water-wheel pumps were used in Paris and London.
	Star-shaped earthen fortifications were developed to resist cannon fire.
ca. 1530	First known horse-powered railway was constructed.
1556	Georgius Agricola published *De Re Metallica,* which described mining methods, ore distribution, pumps, hoists, fans for mine ventilation, mining law, mine surveying, ore processing, and the manufacture of salt, soda, alum, vitriol, sulfur, bitumen, and glass.
ca. 1600	Edmund Gunter developed the "graphicall logarithmic scale," forerunner to the slide rule.
1619	Dud Dudley developed a process to convert coal into coke for cast iron production. Coke replaced charcoal, which was no longer available because forests were decimated.
1642	France's 19-year-old Blaise Pascal devised an adding machine consisting of 10 numbered wheels linked by gears.
1671	Germany's 25-year-old Gottfried von Leibnitz improved upon Pascal's adding machine.
1672	Engineers organized as a separate unit, *Corps du génie*, in the French army.
1698	Britain's Thomas Savery developed the first practical steam engine used to pump water from mines.
1705	Britain's Thomas Newcomen improved the steam engine used to pump water from mines.
1716	French highway department, the *Corps des Ponts et Chaussées*, was organized.
1733	Fly-shuttle loom invented by John Kay in England.
1740	Sulfuric acid production began in England.
1742	America's Benjamin Franklin invented the "Franklin stove," which used fuel more efficiently than fireplaces.
1752	Benjamin Franklin established the similarity between lightning and static electricity in his famous kite-flying experiment.
1759	John Smeaton completed the Eddystone Lighthouse on the treacherous Eddystone Rocks of the English Channel, 14 miles from shore.
1763	Cugnot built a steam locomotive in France.
1770	"Spinning Jenny" was invented by James Hargreaves for making yarn in England.
1775	France's Nicolas LeBlanc developed a process to convert ordinary salt (sodium chloride) to soda (sodium carbonate) for use in glass and soap manufacture. This is regarded as an important milestone in the development of the chemical industry.
1776	First steam engine by Watt and Boulton was installed as a mine pump in England.
1779	First all-metal, cast-iron bridge was constructed in Coalbrookdale, England.
1783	France's Montgolfier brothers flew in hot-air balloons.
1784	First large-scale use of steam for industry was located at Albion Mills in England.
1785	England's Edward Cartwright invented the mechanical loom.
1788	England's William Symington built the first steam-powered boat.

1792	First American canal, only 5 miles long, opened in South Hadley, Massachusetts.
1796	America's Eli Whitney invented the cotton gin for separating cotton from seeds, hulls, etc.
1794	Eli Whitney demonstrated a manufacturing technique based upon interchangeable parts, rather than custom-fitted parts.
ca. 1800	Italy's Volta developed the first battery.
1801	Britain's Sir Humphry Davy developed the electric arc light.
1812	England's 20-year-old Charles Babbage conceived the mechanical "Difference Engine," a calculating machine.
1817	Britain's Henry Cort developed the "puddling process" to transform cast iron to wrought iron.
1818	British Institute of Civil Engineers was founded.
ca. 1820	Analytical mechanics and materials testing were first used for bridge building.
1824	Portland cement, made from lime and clay, was patented by Joseph Aspdin in Britain. This cement improved upon lime mortar known to the ancients, the Greek mixture of lime and Santorin earth, and the Roman mixture of lime and pozzuolana (volcanic ash).
1825	Erie Canal, 363 miles long, joined the Hudson River and Great Lakes.
1829	Britain's George Stephenson built a locomotive called the "Rocket," so named because it could travel at a record speed of 35 miles per hour.
ca. 1830	Britain's William Sturgeon and America's Joseph Henry showed that a magnet is produced when electric current passes through a coiled wire surrounding a metal core.
1831	Britain's Michael Faraday showed that an electric current is induced in a wire when it moves through a magnetic field.
1833	First practical internal combustion engine was developed in England.
	England's Charles Babbage designed the Analytical Engine, the first universal digital computer. It was designed to be programmed using punch cards and could perform logical and arithmetic operations. Unfortunately, it was not built.
1834	America's Cyrus Hall McCormick patented the reaper for harvesting grain.
1836	America's Colt invented the revolver.
1837	Britain's William Cooke and Charles Wheatstone communicated by electric telegraph.
1838	The first Atlantic crossing was made using steam power exclusively. The trip required 15 to 18 days.
1839	Charles Goodyear "vulcanized" rubber by heating rubber latex with sulfur.
1840	There were two engineering schools in the United States.
	Sir William Groves demonstrated an incandescent light by flowing electricity through a platinum wire, but it soon burned out.
1842	The first underwater tunnel was constructed under the Thames River.
1843	America's Samuel Morse commercialized the electric telegraph and sent the first message between Washington and Baltimore.
1845	Guncotton explosive (i.e., cotton treated with nitric and sulfuric acid) was invented. It is more explosive than black powder made of saltpeter (potassium nitrate), sulfur, and charcoal.
1846	America's Elias Howe patented the sewing machine.
	Britain's William Thompson invented pneumatic tires.
1847	James Young patented oil refining by distillation.
ca. 1850	One of the first modern sewage systems was built in Hamburg, Germany.
1852	Henri Giffard powered a dirigible with a steam engine.
1856	Britain's Henry Bessemer invented a steel-making process that allowed steel to be widely produced and ultimately replaced cast and wrought iron in many applications.
1859	The first elevator was developed. It used a steam-powered screw to raise passengers as high as six stories. This invention made skyscrapers possible.
	Edwin Drake's 69-ft-deep oil well came into production, establishing the modern U.S. petroleum industry.
1860	France's Jean-Joseph-Étienne Lenoir made the first practical internal combustion engine.
1861	France's François Coignet demonstrated reinforced concrete by embedding metal bars in *concrete* (Portland cement + sand + stone aggregate), thus improving upon a technology first employed by the ancient Greeks.
1865	Telegraph cable was laid across the Atlantic Ocean, establishing instant communications between America and Europe.
1866	Sweden's Alfred Nobel invented *dynamite* (a mixture of nitroglycerin and diatomaceous earth), an explosive that is safe to handle.
1868	A compressed-air refrigeration plant was built in Paris.
1869	Suez Canal opened.
1870	There were 70 engineering schools in the United States.
ca. 1870	The electric generator was developed using numerous worldwide improvements.
1872	America's John Hyatt opened a factory that produced *celluloid* (guncotton treated with camphor and alcohol), one of the first plastics.

1873   America's Brayton demonstrated an engine that evolved into the jet engine.
        Germany's Carl von Linde developed the first practical ammonia refrigeration machine.
1876   America's Alexander Graham Bell exhibited the telephone at the Philadelphia Centennial.
        Germany's Nikolaus Otto perfected the four-stroke internal combustion engine.
1877   America's Thomas Edison invented the phonograph.
1878   First all-steel bridge was constructed in the United States.
1879   First commercial electric railway was constructed in Berlin.
        First electric power station was installed in San Francisco to power arc lights.
        Thomas Edison invented the lightbulb using a carbonized thread in an evacuated bulb. It lasted almost 2 days.
1882   Thomas Edison started operating the world's first electric generator/electric light system (750 kW) in New York City.
        Von Schroder developed first blood oxygenator machine.
1883   John Roebling's Brooklyn Bridge was completed in New York.
        Sweden's Karl Gustaf Patrick de Laval developed the first practical turbine.
1884   The first American skyscraper (10 stories) was erected in Chicago.
        France's Count Hilaire de Chardonnet patented artificial silk made from nitrated cotton, an explosive.
1885   Germany's Karl Benz built a motorized tricycle.
ca. 1885 America's Frederick Taylor introduced "scientific management" to improve industrial efficiency.
1886   America's Charles Hall developed an electrolytic process to produce aluminum.
1887   Germany's Gottlieb Daimler ran the first motor car.
1888   German physicist Heinrich Hertz built an oscillating circuit that transmitted an electromagnetic wave that induced current in a nearby antenna.
        Alexandre Eiffel constructed the Eiffel Tower in Paris.
        Nikola Tesla patented a multiphase, alternating current, electric motor.
1891   First automobiles were produced in France and Belgium.
1892   Germany's Rudolph Diesel patented an engine using oil as a fuel, rather than gasoline.
        8,000,000 electric lightbulbs were produced.
1894   The turbine-powered (2300-hp) steamship *Turbinia* was launched.
1895   The first large-scale U.S. water power project was completed near Niagara Falls.
1896   America's Samuel B. Langley flew a large steam-powered airplane model.
        Italian inventor Guglielmo Marconi received a patent on wireless radio.
1898   Count Ferdinand von Zeppelin studied rigid, lighter-than-air aircraft.
1900   An electrolytic process was developed to make caustic soda (sodium hydroxide) and chlorine gas from salt (sodium chloride) at Niagara Falls.
        American and British engineers met in Paris to decide whether to standardize on alternating current (AC) or direct current (DC). AC was selected because it can be easily transformed to high voltages for more efficient transmission.
1901   Peter Hewitt developed the mercury-vapor arc lamp, which evolved to the fluorescent tube about 35 years later.
1903   The Wright brothers demonstrated powered flight. In the best flight that year, the plane traveled 852 feet in slightly less than 1 minute.
        Oil-insulated 60,000-volt transformers were developed for efficient electricity transmission.
1904   New York Subway opened.
1905   Albert Einstein proposed the relativity theory, which concluded that $E = mc^2$, i.e., that mass and energy are interchangeable.
1906   The tungsten light filament was introduced, improving lightbulb output 4.7 times and life 27 times.
1907   America's Lee De Forest created the thermionic vacuum tube, the forerunner to the transistor.
1910   Bakelite plastic became a commercial product. It replaced wood, glass, and rubber in many products.
ca. 1910 Germany used the Haber process to fix nitrogen from the air. Although originally used to make explosives for World War I, this process is now used to make fertilizers.
1913   Henry Ford adapted the moving assembly line to automobile production.
1914   The first ship passed through the Panama Canal.
        Robert Goddard began his rocket studies.
1915   X-rays first used for medical imaging.
1920   Spain's Juan de Cierva added an unpowered, horizontal "propeller" to a small-wing airplane to prevent stalling. This was the precursor to the helicopter.
1922   Commercial radio broadcasting was initiated in the United States.
1923   Highly efficient transmission voltages of 220,000 V were used in the western United States.

1925	Electric-powered home refrigerators that used chlorofluorocarbon ("Freon") refrigerants were commercially available. Vannevar Bush constructed the "Differential Analyzer," the first analog computer that mechanically solved sets of differential equations.
1927	The first experimental television was demonstrated by transmitting images from Washington to Bell Laboratories in New York.
1930	Empire State Building (102 stories) was completed.
1931	George Washington Bridge was completed.
1932	Britain's John Cockcroft and E. T. S. Walton confirmed Einstein's theory by bombarding lithium with high-energy protons and measuring the resulting changes in mass and energy.
1934	American chemist Wallace Carothers invented "Nylon 66." Italy's Enrico Fermi bombarded uranium with neutrons and apparently created a new heavier element called neptunium. DeBakey developed the roller pump, which was later used in heart-lung machines.
1936	France's Eugene Houdry developed catalytic oil cracking.
1937	Golden Gate Bridge was completed.
1938	The first commercial fluorescent tubes were sold by General Electric. Germany's Otto Hahn and F. Strassman split uranium by bombardment.
1944	Howard Aiken's "Harvard Mark I" electromechanical computer was built by IBM. It performed 200 additions per minute and worked to 23 significant figures.
1945	The United States detonated the first atomic explosion in Alamogordo, New Mexico. It was the result of the 4-year, $2 billion Manhattan Project. The ENIAC all-electronic computer was completed at the University of Pennsylvania. It used 18,000 vacuum tubes and 6000 switches to perform 5000 additions per second.
1946	Willem Kolff developed the first artificial kidney machine.
1948	The first transistor was demonstrated at Bell Laboratories.
1952	The United States detonated the first fusion bomb at Eniwetok.
1953	First clinical use of the heart-lung machine.
1956	The first full-scale nuclear power plant was completed at Calder Hall, England.
1957	The Soviet Union launched the Sputnik satellite.
1960	Theodore Maiman demonstrated a laser. America's Wilson Greatbatch developed the implantable heart pacemaker.
1961	The Soviet Union placed Yuri Gagarin in orbit.
1969	America's Neil Armstrong walked on the moon. Denton Cooley implanted an artificial heart in a patient.
1975	First CAT scanner developed for medical imaging.
1981	The first U.S. space shuttle was launched. IBM introduced its first personal computer.
1982	Compact discs were first used to store music.
1994	The "Chunnel" was completed linking England with France by tunneling under the English Channel.
1998	The first components of the International Space Station were launched into orbit.
2000	Expedition One of the International Space Station launched. "Rough draft" of the human genome finished.
2004	SpaceShipOne captures Ansari X Prize.

Note: "ca." is an abbreviation for "circa," meaning "about."

INDEX